Engineering Applications of Computational Methods

Volume 23

The book series Engineering Applications of Computational Methods addresses the numerous applications of mathematical theory and the latest computational or numerical methods in various fields of engineering. It emphasizes the practical application of these methods, with possible aspects in programming. New and developing computational methods using big data, machine learning and AI are discussed in this book series, and could be applied to engineering fields, such as manufacturing, industrial engineering, control engineering, civil engineering, energy engineering and material engineering.

The book series Engineering Applications of Computational Methods aims to introduce important computational methods adopted in different engineering projects to researchers and engineers. The individual book volumes in the series are thematic. The goal of each volume is to give readers a comprehensive overview of how the computational methods in a certain engineering area can be used. As a collection, the series provides valuable resources to a wide audience in academia, the engineering research community, industry and anyone else who are looking to expand their knowledge of computational methods.

This book series is indexed in both the **Scopus** and **Compendex** databases.

Hui Liu • Yanfei Li • Zhu Duan

Data Science in Air Quality Monitoring

Springer

Hui Liu
School of Traffic and Transportation Engineering
Central South University
Changsha, Hunan, China

Yanfei Li
School of Mechatronic Engineering
Hunan Agricultural University
Changsha, Hunan, China

Zhu Duan
School of Traffic and Transportation Engineering
Central South University
Changsha, Hunan, China

ISSN 2662-3366 ISSN 2662-3374 (electronic)
Engineering Applications of Computational Methods
ISBN 978-981-96-5776-6 ISBN 978-981-96-5777-3 (eBook)
https://doi.org/10.1007/978-981-96-5777-3

Jointly published with Science Press

The print edition is not for sale in China mainland. Customers from China mainland please order the print book from: Science Press.

Jointly published with Science Press, China

This Springer imprint is published by the registered company Springer Nature Singapore Pte Ltd.
The registered company address is: 152 Beach Road, #21-01/04 Gateway East, Singapore 189721, Singapore

Preface

As globalization and urbanization continue to accelerate, air pollution has become one of the major challenges affecting human health, the environment, and social development. The rapid expansion of industrialization and transportation has led to a large amount of pollutant emissions, making the trend of air quality deterioration more obvious worldwide. Pollutants such as particulate matter and harmful gases in the air not only pose a direct threat to the human respiratory and cardiovascular systems, but also cause irreversible damage to the ecological environment through climate change, acid rain, and other means. In addition, air pollution also brings hidden costs to economic development, including reduced labor productivity due to health problems, increased medical costs, and the cost of resource depletion and ecological restoration.

Air quality monitoring, as an important means of environmental management and public health protection, has received unprecedented attention. With the advancement of science and technology and the surge in data volume, the application of data science and technology in air quality monitoring has gradually shown great potential. The integration of emerging technologies such as big data, the Internet of Things, artificial intelligence, and computational intelligence has enabled air quality monitoring to gradually develop in the direction of digitalization and intelligence.

It is in this context that this book innovatively combines computational intelligence with the specific application of air quality monitoring, focusing on the most cutting-edge technologies and methods in the field of data science, and promoting air quality monitoring into a more intelligent and accurate data-driven era. This book analyzes the application of data science in air quality monitoring in detail; explores the importance of data science in this field; introduces the key impact of air quality monitoring on public health, environment, and economy; and points out the important role of data science. Focusing on the application of data science in air quality monitoring, this book systematically introduces a series of advanced technologies from data preprocessing to data decomposition, identification, clustering, prediction, and interpolation. Each chapter not only combines the latest research results, but also demonstrates the actual effects and applications of various technologies through performance comparison and case analysis. Using advanced methods such as machine learning and deep learning, this book explores how to realize automated air quality monitoring and decision support systems and demonstrates the great potential of these technologies in practical applications through a large number of case analyses. This book provides valuable technical references for researchers in the fields of environmental monitoring, data science, artificial intelligence, etc., and also provides operational theoretical support for policy makers and environmental managers.

In the process of writing the book, the team members have done a lot of experiments and other work, the authors would like to express heartfelt appreciations.

Changsha, Hunan, China

Hui Liu
Yanfei Li
Zhu Duan

Contents

Nomenclature

A	Approximation coefficients
AE	Auto-encoder
AI	Artificial intelligence
ANN	Artificial neural networks
ANNs	Artificial neural networks
AQHI	Air quality health index
AQI	Air quality index
ARIMA	Autoregressive integrated moving average
BiLSTM	Bidirectional long short-term memory
BMU	Best matching unit
BNN	Bayesian neural networks
CAE	Convolutional autoencoder
CEEMD	Complete ensemble empirical mode decomposition
CNNs	Convolutional neural networks
CO	Carbon monoxide
CWT	Continuous wavelet transform
D	Detail coefficients
DBN	Daubechies wavelet
DBSCAN	Density-based spatial clustering of applications with noise
DELM	Deep extreme learning machine
DIE	Dispersion index of error
DTW	Dynamic time warping
DWT	Discrete wavelet transform
EEMD	Ensemble empirical mode decomposition
ELM	Extreme learning machine
EM	Expectation-maximization
EMD	Empirical mode decomposition
EPLS	Ensemble empirical mode decomposition, principal component analysis, and least squares
ESN	Echo state network
EWT	Empirical wavelet transform

GAE	Graph auto-encoders
GARCH	Generalized autoregressive conditional heteroskedasticity
GMM	Gaussian mixture model
GRU	Gated recurrent units
IDW	Inverse distance weighted interpolation
IDWT	Inverse discrete wavelet transform
IMF	Intrinsic mode function
IWE	Interval width of error
KGE	Kling–Gupta efficiency
Kmeans	K-means clustering
KNN	k-nearest neighbors
KSI	Kriging based sequence interpolation
LDA	Linear discriminant analysis
LMD	Local mean decomposition
LSTM	Long short-term memory
MAD	Median absolute deviation
MAE	Mean absolute error
MAPE	Mean absolute percentage error
MAR	Missing at random
MCAR	Missing completely at random
MLP	Multi-layer perceptron
MNAR	Missing not at random
MRA	Multi-resolution analysis
MSE	Mean squared error
NAQFC	National air quality forecast capability
NLP	Natural language processing
NNI	Nearest neighbor interpolation
NO_2	Nitrogen dioxide
O_3	Ozone
OK	Ordinary Kriging
P	Pearson correlation coefficient
PCA	Principal component analysis
PM	Particulate matter
PM10	Particulate matter 2.5 micrometers or less in diameter
PM2.5	Particulate matter 10 micrometers or less in diameter
PMFG	Planar maximally filtered graph
QRNN	Quantile recurrent neural networks
RBF	Radial basis function
RFE	Recursive feature elimination
RMSE	Root Mean Squared Error
RNN	Recurrent neural networks
RNNs	Recurrent neural networks
SD	Standard deviation
SDE	Standard deviation of error
SNR	Signal-to-noise ratio

SO_2	Sulfur dioxide
SOM	Self-organizing map
SSVD	Sparse singular value decomposition
STFT	Short-time Fourier transform
SVD	Singular value decomposition
SVM	Support vector machine
SVR	Support vector regression
TS	Time series
UK	Universal Kriging
VAR	Vector autoregression
VMD	Variational mode decomposition
VOCs	Volatile organic compounds
WD	Wavelet decomposition
WHO	World Health Organization
WPD	Wavelet packet decomposition
XGBoost	eXtreme gradient boosting

List of Figures

List of Tables

Chapter 1
Introduction

Abstract This chapter highlights the importance of air quality monitoring to public health, environmental sustainability, and economic development, as well as its key role in environmental protection and policymaking. It further examines the current landscape of data science in environmental monitoring, acknowledging the inherent complexity of air quality data and the challenges associated with its collection. Based on this, this chapter outlines the current application of data science in air quality monitoring and points out the achievements and shortcomings of existing research. This chapter aims to explore this field in more depth by focusing on the key issues faced by data science in air quality monitoring, including data preprocessing, data decomposition, data identification, data clustering, data prediction, and data interpolation. For each key issue, this chapter discusses its role and importance in air quality monitoring, analyzes the challenges each faces, and outlines the commonly used methods. Finally, this chapter briefly introduces the scope of this book, laying the foundation for the content of subsequent chapters.
This chapter highlights the importance of air quality monitoring to public health, environmental sustainability, and economic development, as well as its key role in environmental protection and policymaking. It further examines the current landscape of data science in environmental monitoring, acknowledging the inherent complexity of air quality data and the challenges associated with its collection. Based on this, this chapter outlines the current application of data science in air quality monitoring and points out the achievements and shortcomings of existing research. This chapter aims to explore this field in more depth by focusing on the key issues faced by data science in air quality monitoring, including data preprocessing, data decomposition, data identification, data clustering, data prediction, and data interpolation. For each key issue, this chapter discusses its role and importance in air quality monitoring, analyzes the challenges each faces, and outlines the commonly used methods. Finally, this chapter briefly introduces the scope of this book, laying the foundation for the content of subsequent chapters.

H. Liu et al., *Data Science in Air Quality Monitoring*, Engineering Applications of Computational Methods 23, https://doi.org/10.1007/978-981-96-5777-3_1

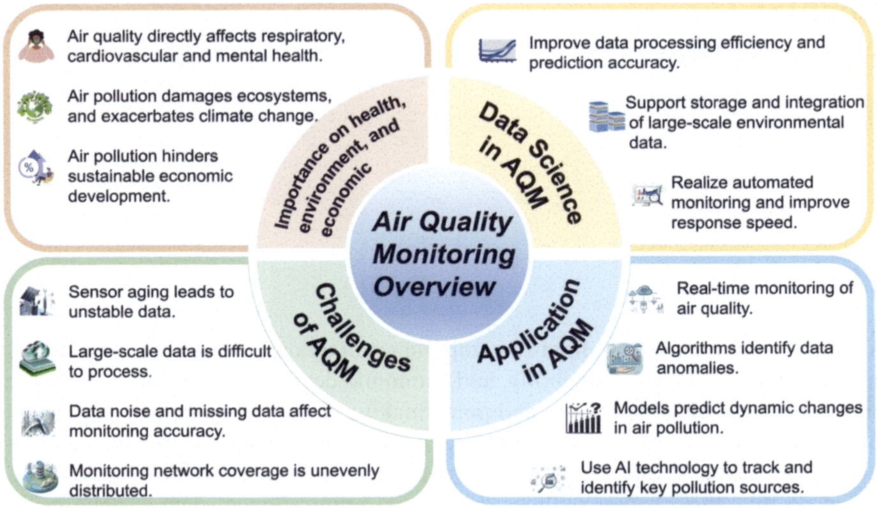

Fig. 1.1 Framework of data science in Air Quality Monitoring (AQM)

1.1 Overview of Data Science in Air Quality Monitoring

Throughout the development of environmental protection, air quality monitoring and management have always been an indispensable part due to their significant implications for public health, environmental sustainability, economic stability, and policymaking. Figure 1.1 shows the framework of data science in air quality monitoring. As air pollution poses severe risks, effective monitoring systems are essential to track, analyze, and manage air quality. Data science and technology have transformative potential to improve the accuracy and reliability of air quality monitoring. This overview provides an in-depth analysis of the importance, challenges, and applications of data science and technology in air quality monitoring.

1.1.1 Importance of Air Quality Monitoring

Since the Industrial Revolution, air pollution has become a major problem threatening human health and environmental sustainability, especially for children. According to the data from the World Health Organization (WHO), in 2019, 99% of the world's population lived in areas where air quality did not meet WHO guidelines (https://www.who.int/news-room/fact-sheets/detail/ambient-(outdoor)-air-quality-and-health). Air quality monitoring is a basic tool for addressing this global challenge and is essential for reducing the negative impact of air pollution. With the advancement of technology, air quality monitoring continues to integrate new technologies such as data science, machine learning, and the Internet of Things, thereby improving its accuracy and efficiency. The importance of air quality monitoring has thus become increasingly prominent and has become a key area in environmental science and policy research.

1.1.1.1 Impact of Air Quality on Public Health, Environment, and Economic Development

Air quality has a far-reaching impact on public health, the environment, and economic development. Air pollution is considered one of the world's most serious public health problems. In areas with rapid urbanization, its threat to human health is particularly significant. Air pollutants, such as fine Particulate Matter (PM2.5, PM10), sulfur dioxide, ozone, and nitrogen oxides, not only cause respiratory diseases but also aggravate a variety of health problems such as cardiovascular disease, asthma, chronic obstructive pulmonary disease, and lung cancer, especially for children, the elderly and people with chronic diseases. Studies have shown that long-term exposure to high concentrations of air pollution will significantly increase the incidence of premature death and chronic diseases, leading to premature deaths and severe health burdens for millions of people worldwide. In addition, air pollution is closely related to mental health problems, especially mental health problems such as depression and anxiety (Kurt et al. 2016). What's more serious is that due to the limitations of indoor ventilation performance and efficiency, the concentration of outdoor pollutants will also affect indoor air quality. In highly polluted weather or urban environments, outdoor air pollutants will enter the room through ventilation systems, door and window gaps, etc., causing a significant drop in indoor air quality.

Air pollution not only harms health but also causes serious damage to the environment. The long-term accumulation of pollutants can lead to ecosystem degradation, affect plant growth, damage the quality of water and soil, further threaten biodiversity, and aggravate the impact of climate change. Rapid industrialization and urbanization have exacerbated this environmental deterioration, especially in developing countries, where air pollution is particularly prominent (Chen and Kan 2008).

Climate change, in turn, will make the air pollution problem more complicated. This complex interaction makes climate change and air pollution problems intertwined, forming a vicious cycle. The increase in greenhouse gases such as carbon dioxide and methane not only directly leads to rising global temperatures, but also changes the chemical reaction process in the atmosphere, further exacerbating the generation of pollutants such as ozone and fine particulate matter.

In terms of the economy, the impact of air pollution cannot be ignored either. Air pollution increases medical costs, reduces labor productivity, and hinders economic development. Some studies have shown that the burden of air pollution on the economy is reflected in work absences, rising medical expenses, and loss of productivity due to health problems. In some countries, the welfare costs of air pollution amount to hundreds of billions of dollars each year, especially in fast-growing countries such as China, where the contradiction between economic development and air quality is particularly prominent (Owusu and Sarkodie 2020).

The continuous occurrence of extreme smog weather caused by air pollution will harm transportation and industrial production. In severe smog weather, the sharp drop in visibility will cause serious obstruction to the operation of the transportation system. For manufacturing and export-oriented economies that rely on efficient transportation systems, frequent smog weather will lead to production delays and increased logistics costs, thereby weakening the market competitiveness of enterprises and increasing economic losses. Smog will also force governments and enterprises to take emergency measures, such as temporary suspension of work or production restrictions, especially in energy-intensive industries such as steel, chemicals, and construction. Especially in areas with more serious pollution, companies may face higher environmental supervision fees and tax pressure, which will further weaken their profitability.

In addition, frequent smog weather will also affect industries that rely on outdoor activities, such as tourism, retail, and services. The main impacts of air quality can be summarized as shown in Fig. 1.2. In the long run, this economic loss caused by air pollution will further affect the sustainable development of the local economy, hinder the diversification of the regional economy, and the improvement of innovation capabilities.

1.1.1.2 Air Quality Monitoring Plays a Key Role in Environmental Protection and Policymaking

As countries worldwide pay more and more attention to environmental protection, the strategic position of air quality monitoring in policymaking continues to rise. As air pollution problems intensify, scientific and accurate air quality monitoring

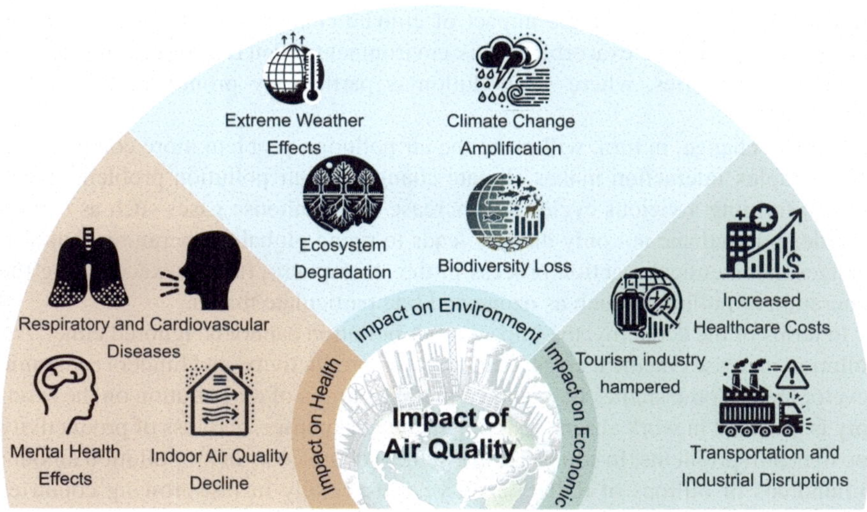

Fig. 1.2 Summary of Air Quality Impact

has become an important means for countries to promote environmental protection strategies. Air quality monitoring not only provides timely data support for the protection of public health and ecosystems but also provides a basis for economic decision-makers to evaluate the effectiveness of pollution control measures (Krzyzanowski et al. 2005). For example, in policy evaluation, monitoring data are widely used to track changes in pollutant emissions in various regions to determine whether air quality standards have been met, and also to help identify which regions or industries have the most effective pollution control measures (Bell et al. 2011).

The real-time and efficient nature of air quality monitoring enables the government to respond quickly when air quality problems arise. The early warning mechanism based on the monitoring system can not only take emergency measures promptly when pollution incidents occur but also predict future pollution trends by analyzing historical data, thereby formulating more refined long-term plans. Taking China as an example, policymakers use air quality monitoring data as a key indicator to measure the effect of economic transformation and ensure that economic development and environmental protection advance simultaneously.

Air quality monitoring also plays a vital role in policymaking in the transportation sector. As urbanization accelerates, transportation has an increasing impact on air quality. In the process of promoting green transportation, air quality monitoring data helps the government measure the actual effect of policies, such as reducing pollution emissions in the transportation sector by increasing the use of new energy vehicles and optimizing public transportation systems.

In terms of economic development, the role of air quality monitoring in promoting the green economy is gradually emerging. Through monitoring data, the government can assess the impact of pollution on economic activities, thereby prompting high-pollution industries to accelerate green transformation. The continuous improvement of air quality can not only improve public health and reduce medical expenses, but also increase labor productivity, thereby providing impetus for the sustainable development of the economy.

In general, air quality monitoring provides technical support for environmental protection and pollution prevention and plays a fundamental role in the sustainable development of the green economy and society. It helps policymakers find a balance between environmental protection and economic development through scientific and precise data support and plays a key supporting role in addressing climate change, improving public health, and promoting green economic transformation. With the continuous advancement of technology, air quality monitoring will continue to help the effective implementation of environmental protection policies, promote the green and low-carbon transformation of the economy and society, and achieve the coordinated development of the environment and the economy.

1.1.2 *The Role of Data Science in Environmental Monitoring*

1.1.2.1 The Development of Data Science and Its Application in Different Fields

The rapid development of data science has penetrated many fields and has brought far-reaching impacts. Figure 1.3 shows the application of data science in some key fields, from business and finance to environmental science, healthcare, engineering technology, and education. Data science has changed the traditional workflow and provided new solutions and unprecedented insights for various fields.

Data science is driving the transformation from empirical decision-making to data-driven decision-making in the business and financial fields. By analyzing large amounts of financial data, data science has significantly improved the accuracy of risk management, consumer behavior analysis, and market forecasting. Through

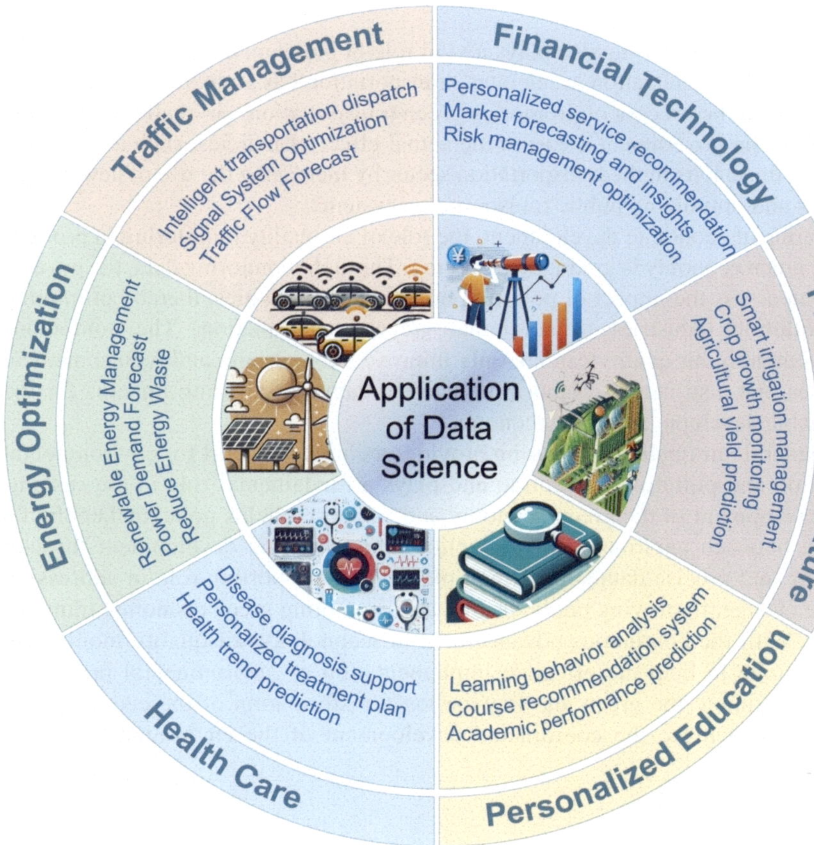

Fig. 1.3 Application of data science in different fields

personalized service recommendations such as credit scoring and insurance pricing, the application of data science has successfully helped banks and insurance companies optimize operations and improve customer experience. This data-driven financial innovation has provided important support for the rapid development of the financial technology field and has also made the financial services industry more intelligent and efficient (Giudici 2018).

The application of data science healthcare is also outstanding. By analyzing electronic health records, large-scale genetic data, and clinical trial data, data science has not only improved the accuracy of disease diagnosis but also accelerated the development of new drugs. Machine learning and deep learning technologies help medical institutions predict the development trend of patients' conditions and support the formulation of personalized treatment plans. For example, predictive analysis based on biomedical data can help doctors better judge patients' recovery speed or possible complications, greatly improving the efficiency of medical resource allocation (Saracco 2020). In addition, with the development of data science, researchers in the medical field can use data integration and analysis techniques to better understand the molecular mechanisms of diseases, thereby finding new treatment pathways for complex diseases such as cancer and diabetes. In the agricultural field, data science helps farmers improve production efficiency and optimize resource utilization by analyzing climate, soil, and crop growth data. Data-driven agricultural technologies, including precision agriculture and smart irrigation, can optimize planting decisions through real-time monitoring and prediction, reduce water and fertilizer waste, and increase food production. In particular, through the combination of remote sensing data and drone technology, farmers can obtain more detailed field data to better manage the crop growth process. In addition, data science is also used to study the dynamic changes in the global food supply chain and help governments and non-governmental organizations formulate policies and strategies to respond to food crises (Blair et al. 2019).

In the field of social sciences, data science provides a new way to study human behavior and social dynamics. Through the analysis of social media, survey data, and other social behavior data, researchers can identify social trends, changes in public sentiment, and policy effects. For example, social scientists can use big data analysis tools to study the public's response to specific policies or events, thereby helping the government improve public policy formulation.

In the field of energy management, data science helps power companies optimize energy distribution and improve efficiency by analyzing energy usage patterns. For example, by analyzing smart meter data, power companies can predict peak hours of electricity demand and adjust the load of the power grid according to actual demand, thereby reducing energy waste. In addition, data science can also be used for the optimal management of renewable energy, such as performance monitoring and prediction of wind and solar systems. These data-driven technologies help energy companies better manage resources and provide technical support for promoting the widespread application of sustainable energy (Rasool and Chaudhary 2022). As global energy demand continues to grow, data science will play a more important role in energy optimization, environmental protection, and addressing climate change.

The field of environmental science has also benefited greatly from the intervention of data science. In response to climate change and environmental protection, data science provides powerful analytical tools for processing and analyzing massive amounts of data from global monitoring networks. Data science can not only accurately predict climate patterns, but also help scientists understand changes in ecosystems and evaluate the impact of policy measures on the environment. For example, by analyzing complex meteorological and environmental data, researchers can more accurately predict the occurrence of extreme weather events and natural disasters, thereby providing more reliable data support for governments and decision-makers (Blair et al. 2019).

In the fields of engineering and industry, data science is promoting the process of automation and intelligence. For example, in traffic management and urban planning, data science techniques are used to predict traffic flow, optimize traffic signal systems, and reduce congestion. In construction projects, data science can help predict the risks that may be encountered during construction, reduce waste, and improve production efficiency by optimizing resource allocation (Rasool and Chaudhary 2022). In addition, the manufacturing industry optimizes production processes through data analysis, improving product quality and production efficiency.

The application of data science in education and academia is mainly focused on student behavior analysis, course recommendations, and personalized learning paths. Data science can not only help educational institutions better understand students' learning patterns but also tailor educational programs according to the needs of each student, thereby improving learning efficiency.

In addition, data science has also had a significant impact on fields such as library information management, structural biology, and computer science. The library field has improved information storage and management methods through data science, providing users with more accurate services (Virkus and Garoufallou 2019).

1.1.2.2 Emphasize the Important Role of Data Science in Environmental Monitoring

Figure 1.4 shows the workflow of data science in environmental monitoring, from data collection, preprocessing, storage, and analysis to final decision support. The application scope of data science in environmental monitoring continues to expand, from processing complex data to improving decision-making efficiency, and has become a core part of the modern environmental monitoring system. With the diversification of environmental data sources and the surge in data volume, traditional data processing methods can no longer meet existing needs. Data science, through its powerful computing power and algorithm optimization, analyzes massive environmental data more efficiently and accurately. For example, environmental monitoring generates massive spatiotemporal data through multi-dimensional sensor

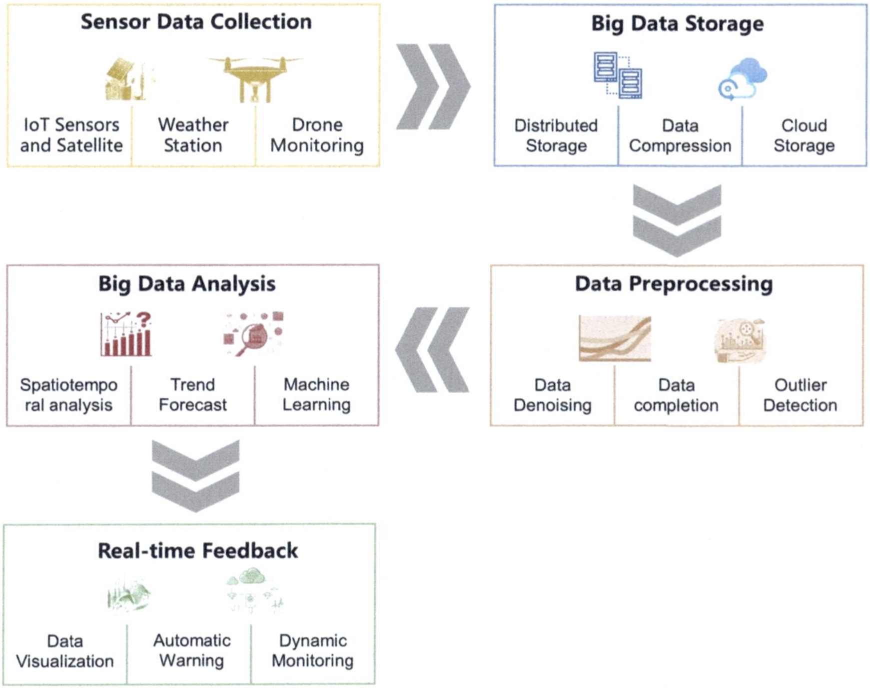

Fig. 1.4 Data Science Workflow in Environmental Monitoring

networks and high-resolution satellite observations, which are difficult to effectively integrate and interpret with traditional methods. Data science techniques, such as big data analysis, machine learning, and statistical modeling, can not only address these challenges but also provide new solutions (Martinez Bilesio et al. 2019). In addition, the application of data science techniques in environmental monitoring has significantly improved the quality control and reliability of data. In complex ecological monitoring systems, data quality is often a key factor in determining the reliability of results. Data science ensures the accuracy of monitoring data through automated status labeling and quality inspection techniques. Through machine learning algorithms, the system can automatically identify abnormal data and ensure the accuracy and availability of data through status labeling technology. This not only reduces human errors but also improves the overall quality of the data. Therefore, data science provides strong quality assurance for environmental monitoring, effectively improving the accuracy of data analysis and the reliability of monitoring results.

The role of data science in environmental monitoring is not limited to data processing, it also provides strong support for environmental system modeling and prediction. Environmental management involves a large number of dynamically changing factors, including climate, pollution, land use, etc., and traditional monitoring methods make it difficult to track these changes in real time. Data science responds to this challenge through complex algorithms and models. Machine learning and statistical methods applied to environmental health research help process high-dimensional pollutant data, identify key health risk factors, and improve the accuracy and efficiency of prediction models (Choirat et al. 2019). In addition, prediction systems based on spatiotemporal modeling, with the support of data science technology, can better explain environmental trends and provide decision-makers with reliable future forecasts (Burr et al. 2023; Zammit-Mangion 2023). This ability to combine big data and predictive modeling enables environmental monitoring to not only observe in real time but also predict the occurrence of ecological risks and extreme weather events in advance (Blair et al. 2019).

Another key contribution of data science is that it helps environmental monitoring to be automated and real-time. By integrating Internet of Things (IoT) technology, data science can collect environmental data from different locations in real time and quickly feed it back to the central processing system. Data science technology can quickly identify abnormal changes in the environment and trigger corresponding response measures by processing and analyzing these multi-source data in real time. Especially when monitoring complex environmental phenomena such as climate change and extreme weather events, data science helps monitoring systems have stronger resilience.

The role of data science in environmental monitoring is also reflected in promoting public participation and transparency. Through the popularization of digital data entry and big data platforms, more and more citizen scientists can participate in the collection and analysis of environmental data. Data science technology makes participatory environmental monitoring more feasible. The public can participate in data collection through simple mobile devices, which not only expands the coverage of data but also increases the breadth of monitoring. Public participation in monitoring can not only supplement the shortcomings of the official monitoring system but also increase the awareness and attention of the whole society to environmental issues. For example, in public participation monitoring projects, digital data entry technology has been widely used in monitoring projects in different regions and has significantly improved the sustainability of the projects and the accuracy of the data (Brammer et al. 2016; Viqueira et al. 2020).

In short, the role of data science in environmental monitoring is irreplaceable. From data processing to predictive modeling, from automated monitoring to public participation, data science is reshaping the future of modern environmental monitoring in all aspects (Blair et al. 2019; Burr et al. 2023; Manfreda et al. 2018).

1.1.3 Characteristics and Challenges of Air Quality Data

1.1.3.1 Complexity of Air Quality Data

The complexity of air quality data stems from the interweaving of multiple factors such as high dimensionality, nonlinearity, and spatiotemporal correlation. However, its complexity is not only reflected in the high dimensionality of the data but also involves the processing and analysis of a large amount of heterogeneous data. Because there are many types of pollutants monitored for air quality, the concentration changes of each pollutant at different times and spaces are also different, showing complex dynamic characteristics. This multi-dimensional, multi-variable data form requires the use of complex data processing techniques for dimensionality reduction and pattern recognition. The processing of high-dimensional data is not only about the number of variables but also involves the interdependence between these variables. As the scale of data continues to increase, how to reduce redundant information while maintaining the important features of the data has become an important challenge in air quality data analysis. Figure 1.5 provides a visual representation of the key challenges of air quality data, including high dimensionality, nonlinearity, spatiotemporal correlation, data noise, and missing or incomplete data.

The nonlinearity of air quality data also increases the difficulty of its analysis. The concentration of air pollutants is affected by a variety of nonlinear factors, such as meteorological conditions, emission intensity of pollution sources, topography, etc. Because the concentration of pollutants fluctuates complexly over time and space, and the interactions between different pollutants also have nonlinear characteristics, traditional linear models perform poorly in predicting air quality, and more complex models are needed to capture the changes in pollutant concentrations.

Spatiotemporal correlation is another important dimension of the complexity of air quality data. The spatiotemporal variation of air pollution is affected by many factors, which are closely related to geographical location, time, and meteorological conditions in addition to emission sources. The spatiotemporal variation of pollutant concentrations often manifests as a highly complex dynamic process, and this variation may have obvious regional differences and temporal discontinuities. Therefore, it is crucial to understand the interrelationships between pollutants in time and space.

In addition to high dimensionality, nonlinearity, and spatiotemporal correlation, the complexity of air quality data is also reflected in data uncertainty, noise level, diversity of pollution sources, and missing or incomplete data. The existence of

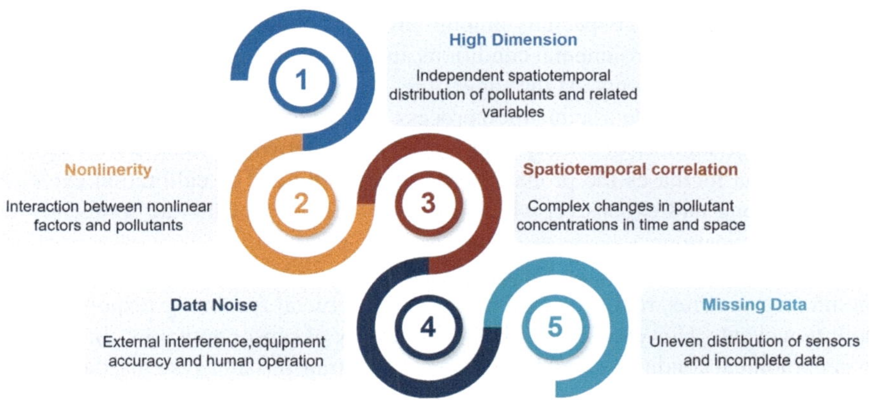

Fig. 1.5 Challenges of air quality data complexity

noise levels also makes air quality data more complex. The noise in air quality data comes from many aspects, including the accuracy limitations of monitoring equipment, interference from environmental factors, and human operation errors. These noises not only affect the accuracy of the data but may also mask the true concentration changes of pollutants, bringing challenges to data analysis and prediction. The complexity of air quality data is also closely related to the diversity of pollution sources. Different pollution sources, such as industrial emissions, traffic pollution, and natural sources (such as sandstorms or forest fires), all have different impacts on air quality. Different pollution sources have their emission characteristics and spatiotemporal distribution patterns, so a single pollution source analysis method often cannot effectively deal with complex air quality data, and there are interactions between different pollution sources, which further increases the difficulty of data analysis. Data uncertainty and missingness are also key issues in the complexity of air quality data. In the actual monitoring process, air quality data often have discontinuities, missing data, or data incompleteness due to equipment failure. This uncertainty poses a huge challenge to air quality prediction and analysis, especially in the processing of long-term series data.

Finally, the complexity of air quality data is also reflected in the cross-scale characteristics of the data. Changes in pollutant concentrations not only have short-term dynamic characteristics but also involve long-term trend analysis. Air pollutants have significant differences at different temporal and spatial scales. Some pollutants may accumulate rapidly in a short period due to changes in meteorological conditions, while they may be affected by seasonal changes on a longer time scale.

1.1.3.2 Difficulties in Data Acquisition

In air quality monitoring, data acquisition and data quality issues affect the accuracy and reliability. Air quality monitoring systems involve multiple complex links, from sensor data acquisition to data transmission, storage, and processing. Each link may be affected by multiple factors, leading to data quality issues.

One of the core issues of air quality monitoring systems is the reliability of equipment and sensors. Especially in long-term continuous monitoring, sensors are easily affected by environmental conditions and equipment aging, which affects the accuracy of measurements. At the same time, sensor failure and data loss are other problems faced in the data acquisition process. Air quality monitoring systems often rely on a large number of widely distributed sensor networks. Long-term continuous operation increases the probability of sensor failure. The calibration cycle of sensors is long and expensive, which makes it difficult to calibrate frequently in practical applications.

Another major challenge is the timeliness and consistency of data. In air quality monitoring systems, real-time data acquisition is crucial for timely response to air pollution events. However, due to the heterogeneity of sensor networks, differences in geographical conditions, and limitations of data transmission systems, data transmission delays or losses often occur.

In addition to the influence of sensors and the environment, the acquisition of air quality data is also limited by the data processing and storage system. Large-scale air quality monitoring systems often generate massive amounts of data, and traditional data processing systems cannot efficiently perform data analysis and inference. Processing platforms based on cloud computing and big data technologies have been introduced into air quality monitoring to meet the needs of large-scale data collection and processing.

Due to cost constraints and a wide geographical range, air quality data collection systems often face the problem of uneven data distribution and scarcity. In low-income countries with limited resources or areas, the high cost and maintenance costs of monitoring equipment limit the coverage of the monitoring network. This data asymmetry is particularly evident in urban and rural areas, where urban centers are usually equipped with more dense monitoring stations, while vast suburbs and remote areas lack sufficient monitoring coverage. Remote sensing technology and low-cost sensors can fill the monitoring gaps and expand the monitoring coverage to a certain extent. However, there are still problems with data quality and accuracy, especially low-cost sensors, which are less accurate and susceptible to environmental influences, and the data generated often need to be strictly post-calibrated and processed (Castell et al. 2017).

1.1.4 Current Application of Data Science and Technology in Air Quality Monitoring

1.1.4.1 Current Application of Data Science and Technology in Air Quality Monitoring

Data science and technology, especially machine learning, deep learning, and big data analysis, have been widely used in the field of air quality monitoring. Figure 1.6 shows an overview of the application of big data analysis, machine learning, and deep learning in air quality monitoring. The application of these technologies has significantly improved the accuracy of air pollution prediction, monitoring efficiency, and the ability to deal with complex pollution sources. With the popularization of sensor networks and the IoT, data science and technology are used to analyze the massive data collected from various environmental sensors and improve the accuracy of pollution monitoring through intelligent algorithms. For example, monitoring systems based on wireless sensor networks and cloud computing use artificial intelligence technology to monitor air quality in real-time through low-power sensors and use cloud platforms for data processing and pollutant identification, so that air quality monitoring systems can work efficiently over a large area and at a relatively low cost (Arroyo et al. 2019). This technology makes air pollution monitoring and data processing timelier and more accurate, especially in rapidly urbanizing areas, and can capture the dynamic changes of pollutants promptly.

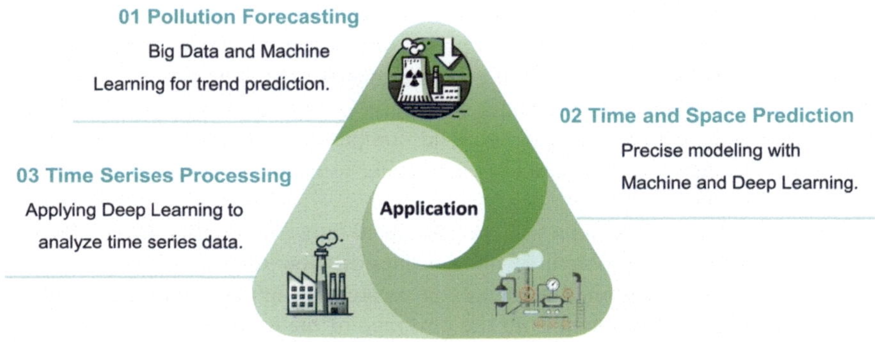

Fig. 1.6 Application of Data Science Technology in Air Quality Monitoring

Machine learning and deep learning technologies have been widely used in air quality monitoring and prediction. Algorithms such as Artificial Neural Networks (ANNs) and Support Vector Machine (SVM) have been shown to have high accuracy and reliability in predicting air pollutant concentrations. In addition, deep learning technology, especially Long Short-Term Memory Network (LSTM), has also been used to process complex time series air quality data. By training models, it is possible to accurately predict changes in air quality while calibrating low-cost monitoring equipment, greatly improving the prediction accuracy (Liu et al. 2020c). These technologies are not only used for pollution prediction but are also widely used for fault detection and anomaly identification in air quality data. In air quality monitoring networks, sensor failures or data anomalies may lead to inaccurate monitoring results, while artificial intelligence algorithms, especially anomaly detection models, can effectively detect and eliminate anomalies in monitoring through data preprocessing, feature selection, and abnormal pattern recognition (Evagelopoulos et al. 2023). In addition, Artificial Intelligence (AI) technology is also used to improve the identification and traceability of air pollution sources. Through AI models, especially artificial neural networks, key pollution sources can be identified from complex pollution source data, thereby providing a scientific basis for decision-makers.

The key advantage of big data technology in air quality monitoring is its ability to process massive data. Traditional data processing methods are generally unable to cope with the massive sensor data in air quality monitoring, while big data technology can efficiently process, analyze, and store large-scale data sets through distributed computing and cloud storage platforms. These data processing technologies not only improve the speed of data analysis but also ensure the integrity and accuracy of the data so that the air quality monitoring system can respond to changes in air pollution more quickly (Arroyo et al. 2019). Through these technologies, air quality monitoring systems can achieve a wider range of monitoring coverage and integrate multiple data sources over a large area to form a more comprehensive monitoring network.

The application of big data technology in air quality monitoring is not limited to data collection and storage. It can also mine hidden patterns and trends through advanced data analysis techniques. Data mining and machine learning algorithms can process large amounts of historical data and use regression analysis, clustering algorithms, and neural networks to analyze the complex relationship between pollutant concentrations and multiple factors such as meteorological conditions, traffic flow, and industrial activities. This multidimensional analysis based on big data helps to reveal the spatiotemporal distribution of air pollutants and their potential causes, thereby providing a scientific basis for policymakers (Shukla et al. 2023). In addition, by integrating multi-source data including meteorological data, pollutant emission data, and traffic and industrial activity data, big data technology can generate sophisticated spatiotemporal prediction models. These models can be updated in real time and continuously optimized with the input of new data to provide more accurate air pollution forecasts. This dynamic prediction capability is particularly important when dealing with extreme air pollution events (such as smog weather). It can help city managers take corresponding control measures promptly to reduce the harm of air pollution to public health. In terms of air pollution control, big data technology also provides powerful data support and decision-making assistance tools. Through big data analysis, decision-makers can evaluate the specific contribution of different pollution sources to air quality and formulate targeted pollution control strategies.

1.1.4.2 Achievements and Shortcomings of Existing Research

Current research shows that the application of data science technology, especially AI, machine learning, and big data analysis, in air quality monitoring has significantly enhanced the accuracy and efficiency of pollution monitoring and prediction. In recent years, the use of AI technology has expanded from simple air quality data processing to complex real-time pollution prediction and automated response systems. For example, a monitoring system combining the IoT and artificial intelligence can collect data in real-time through a low-cost sensor network and use big data technology to predict and warn pollutant concentrations. This approach significantly improves monitoring accuracy. In applications in high-traffic areas, artificial intelligence technology has demonstrated its strong ability to process complex spatiotemporal data (Kulikova et al. 2023). Research also shows that the application of data science technology in air pollution fault detection can effectively reduce the occurrence of data anomalies and errors, thereby improving the overall reliability of the monitoring system. Another application of AI in air quality monitoring is a real-time pollution warning and automated response system. By combining the Internet of Things and machine learning algorithms, the monitoring system can predict future pollution events and actively intervene. For example, based on historical pollution data and real-time meteorological conditions, AI systems can predict air quality in the next few hours and provide timely intervention recommendations to decision-makers (Arroyo et al. 2019). This proactive early warning system can not only help government agencies respond to sudden pollution incidents more effectively but also provide the public with real-time air quality information. However, existing studies have pointed out that the application effect of these systems in large-scale urban environments is still limited by many factors, including the real-time nature of data acquisition, the scalability of the system, and the specific pollution patterns of different cities. Especially in urban areas with complex pollution sources, how to establish an effective causal relationship model between multiple pollution sources remains an urgent problem to be solved.

Although current research results show the great potential of AI and data science technologies in air quality monitoring, these technologies still face challenges in dealing with the complexity of environmental monitoring. First, the diversity and complexity of data make AI models highly dependent on data quality. Although a large amount of data can be effectively collected through sensor networks and cloud computing platforms, these data often contain noise or are incomplete, affecting the accuracy of predictions. Therefore, algorithms that rely on high-quality big data for training and prediction have limitations in practical applications because the data in the sensor network is often limited by the accuracy of the sensors, the frequency of data collection, and the lack of geographical coverage. The model training process relies on a large amount of high-quality labeled data, which limits the generalization ability of the model in air quality monitoring. In some cases, air quality models may perform well in a specific area, but when applied to other areas, the prediction accuracy of the model will drop significantly due to environmental differences such as climate conditions and distribution of pollution sources.

The geographical coverage of sensors is also a major bottleneck. In areas with complex geographical conditions, such as valleys and densely populated areas of urban high-rise buildings, the number and layout of sensors are usually difficult to cover the air quality conditions in all areas. This monitoring gap further exacerbates the locality of the data, making AI models face more uncertainty and bias when predicting air quality changes over a large area.

1.2 Key Problems Data Science in Air Quality Monitoring

1.2.1 Data Processing

Air quality data processing involves multiple aspects of quality issues and technical means. Figure 1.7 shows the main challenges faced in data processing, the key steps of data preprocessing, and common data processing methods. This figure can provide a clearer understanding of the core issues and countermeasures involved in air quality monitoring data processing.

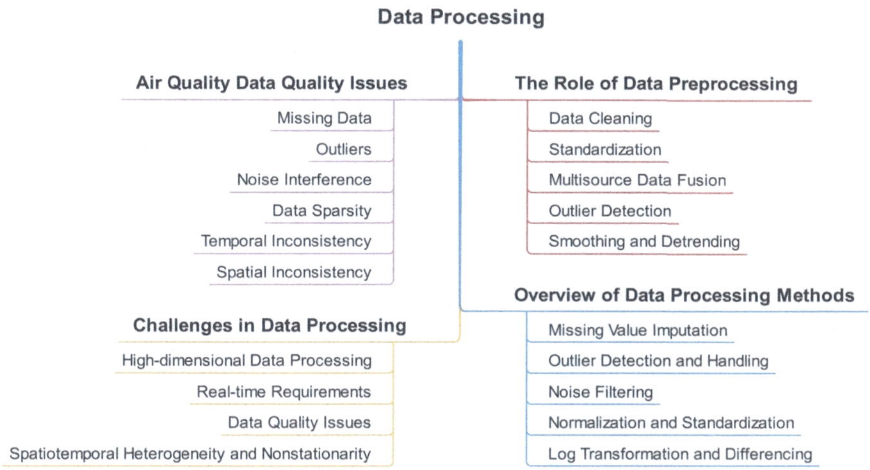

Fig. 1.7 Key issues and technologies in air quality data processing

1.2.1.1 The Role of Data Preprocessing in Air Quality Monitoring

The quality issues in air quality data are one of the important challenges faced in environmental monitoring. These issues include missing values, outliers, and noise interference, which seriously affect the accuracy of data analysis and the reliability of subsequent model construction. Data preprocessing is an important step to ensure data reliability and validity. Data preprocessing ensures a solid foundation for subsequent analysis and model building by cleaning, filtering, and standardizing these data. In the process of air quality monitoring, outlier detection and data cleaning steps can effectively identify data errors caused by sensor failure or environmental interference, and improve the overall reliability of the data. By applying artificial intelligence methods for data preprocessing, not only can data quality be improved, but also problems in the monitoring network can be found through outlier detection, making the data more reliable (Evagelopoulos et al. 2023).

Another key role of data preprocessing is to deal with missing and inconsistent data. Air quality monitoring systems often face sensor failures or communication interruptions, resulting in partial data missing or erroneous transmission. In this case, preprocessing techniques such as data interpolation and standardization can effectively fill data gaps and ensure that data from different sources have consistent formats and units. Preprocessing methods can significantly improve data accuracy and the reliability of model predictions by integrating multi-source data and ensuring data consistency. These techniques are not only applicable to air quality monitoring systems but can also be applied to the prediction and analysis of large-scale data (Simo et al. 2020).

In real-time monitoring systems, the preprocessing step is particularly important because it can ensure rapid response and processing of data.

First, data cleaning is one of the basic steps of preprocessing, which aims to remove noise and outliers in the data. Sensors may generate erroneous data due to external conditions or equipment failures. These data are identified and removed during the cleaning process to ensure the accuracy and consistency of the monitoring data.

Secondly, standardization is an indispensable part of data preprocessing, especially when integrating data from different sources. In addition, multi-source data fusion is an important step in processing data from different monitoring systems. By integrating multi-dimensional data sources, preprocessing can improve the integrity and representativeness of data. For example, when using drones, sensors, and satellite data for air quality monitoring, data fusion can enhance spatial coverage, remove redundant information through feature extraction technology, and retain key data points for model training and analysis (Wivou et al. 2016).

Outlier detection is also an important part of data preprocessing. Outliers may be caused by extreme events or equipment failures. If these data are not processed, they will mislead the analysis results. Using artificial intelligence and machine learning models, such as the isolation forest model, abnormal patterns in air quality data can be automatically identified and these outliers can be removed before model training to reduce the propagation of errors (Evagelopoulos et al. 2023).

Data smoothing and detrending help eliminate random fluctuations in the data, making long-term trends and cyclical changes more obvious. In air quality forecasting, data smoothing can weaken the impact of short-term fluctuations and improve the predictive ability and accuracy of the model. This detrending process makes the monitoring data more stable and improves the predictive ability of the machine learning algorithm. Especially in complex air quality forecasting, this technology can effectively reduce prediction errors (Simo et al. 2020).

1.2.1.2 Main Challenges in Data Processing

In air quality monitoring data processing, high-dimensional data processing is a key challenge. With the continuous advancement of monitoring technology, the diversification of data sources and monitoring indicators has led to a rapid increase in data dimensions, forming a "dimensionality disaster" problem. In addition, high-dimensional data usually contains a large number of redundant and irrelevant features, and these noisy data will further degrade the performance of the model. Xu et al. (2019) proposed to use feature selection and dimensionality reduction techniques to improve the performance of air quality inference models when data is sparse by fusing remote sensing data with urban data. By reducing the dimensionality and optimizing the computational performance of the model, they effectively overcame the challenges brought by data sparsity and high-dimensional data processing (Xu et al. 2019).

Real-time requirements are another important technical difficulty in air quality monitoring. In real-time monitoring, data is usually continuously generated from sensor networks or other monitoring devices in the form of streams, and the system needs to process them with the lowest possible latency. To overcome this challenge, many studies have adopted technologies such as distributed computing and edge computing to improve the real-time processing capabilities of the system. Ameer et al. (2019) compared the real-time processing performance of several different machine learning technologies on large-scale data sets and found that the distributed framework based on Apache Spark can significantly improve processing efficiency and meet the needs of real-time monitoring (Ameer et al. 2019). Wardana et al. (2021) proposed a deep learning model optimized for edge devices, which successfully addressed the real-time challenge under resource-constrained conditions by reducing computational overhead (Wardana et al. 2021).

Among the challenges unique to air quality monitoring, spatiotemporal heterogeneity and non-stationarity are particularly significant. The spatial distribution and temporal variation of air pollution are affected by many factors, such as meteorological conditions, traffic flow, topography, etc., which makes the concentration variation of pollutants highly dynamic and complex. Zhu et al. (2017) proposed a spatiotemporal Granger causality model, which helps improve the accuracy of air quality inference in the case of sparse data by analyzing the causal relationship between urban dynamics and air quality (Zhu et al. 2017).

In summary, data processing in air quality monitoring faces multiple technical difficulties, especially in high-dimensional data processing, real-time requirements, and data quality assurance. At the same time, the spatiotemporal heterogeneity and non-stationarity of air pollution pose unique challenges to air quality monitoring. By optimizing algorithms, using distributed computing and data assimilation techniques, researchers are gradually overcoming these difficulties, improving the accuracy and efficiency of air quality monitoring, and providing reliable data support for environmental protection and health risk assessment.

1.2.1.3 Overview of Data Processing Methods

Data preprocessing is crucial in air quality monitoring data processing. Table 1.1 shows the principles and applicable scenarios of different methods such as data cleaning, outlier processing, noise filtering, and data transformation, indicating the key role played by each method in air quality monitoring data processing.

First of all, data cleaning is an important link, especially the processing of missing values. Common methods for processing missing values include mean or median filling, interpolation, and multiple interpolation. Mean or median filling is the simplest way, replacing missing values with common values in the data set. Although it is easy to implement, it may introduce bias when the data distribution is asymmetric. Interpolation rules infer missing values based on adjacent data, such as linear interpolation and spline interpolation, which is particularly common in time series data and can maintain the continuity of spatial and temporal data. Multiple interpolation is a more complex statistical method that predicts missing values by building a model. It is particularly suitable for complex data sets and can generate multiple

Table 1.1 Overview of data processing methods

Processing method	Principle	Applicable scenarios	Literature support
Missing value filling	Use mean, interpolation, or multiple interpolation to predict and fill missing values	When missing data affects subsequent analysis	Brantley et al. (2014); Miasayedava et al. (2023)
Outlier detection and processing	Use statistical or machine learning methods to identify outliers	When there are outliers or anomalies in the data set	Roosken (1978); Ameer et al. (2019)
Noise filtering	Use Kalman filtering, moving average, and other methods to remove noise in data	When it is necessary to smooth or remove random fluctuations in data	Benammar et al. (2018)
Normalization and standardization	Scale data to a specific range or adjust data to a standard normal distribution	When the numerical ranges of different features vary greatly	Yang et al. (2019)
Logarithmic transformation and difference	Logarithmic transformation to handle skewed distributions, and differences to manage non-stationary time series data	When the data has a seriously skewed distribution or non-stationary time series	Wardana et al. (2021)

different data sets to reduce estimation bias (Brantley et al. 2014; Miasayedava et al. 2023).

Outlier detection is another key step in data cleaning. Traditional statistical methods such as Z-score and boxplot are often used to detect and process outliers, which can effectively identify points far from the center of data distribution. However, with the increase in data complexity, machine learning methods such as isolation forest and Density-Based Spatial Clustering of Applications with Noise (DBSCAN) are widely used. These methods can handle complex high-dimensional data, especially in smart city air quality prediction, and achieve more efficient outlier detection and processing through machine learning algorithms (Roosken 1978; Ameer et al. 2019). In addition, noise filtering is also a necessary step to ensure data quality. Common methods include Kalman filtering and moving average. The former eliminates random noise in the data through state estimation, and the latter reduces fluctuations by smoothing data. These methods are particularly important in real-time data processing and can improve the stability and accuracy of monitoring data (Benammar et al. 2018).

In the process of data transformation, normalization and standardization are the most commonly used techniques. Normalization ensures the numerical comparability between features by scaling the data to a specific range (such as 0–1), which is suitable for models with sensitive input value ranges. Standardization adjusts the data to a standard normal distribution with a mean of 0 and a variance of 1, which is suitable for comparison and analysis of data of different scales. These transformation methods are widely used in machine learning and big data analysis to ensure the balance between features (Yang et al. 2019). In addition, logarithmic transformation and difference are also common means of data transformation. Logarithmic transformation processes skewed distributed data to make it closer to normal distribution, which is suitable for processing data with large differences. Difference methods are used in time series analysis. By converting non-stationary data into stationary series, it helps meet the assumptions of statistical models and improves prediction accuracy (Wardana et al. 2021).

1.2.2 Data Decomposition

Data decomposition plays a vital role in the feature extraction and processing of complex data sets. Figure 1.8 illustrates the main functions of data decomposition in reducing complexity and extracting key features, highlights the challenges that arise when applying these techniques, and provides an overview of decomposition methods. The figure provides a structured perspective on how data decomposition can help in efficient feature extraction and decision-making in environmental data analysis such as air quality monitoring.

1.2.2.1 The Role of Data Decomposition in Feature Extraction

Decomposition methods such as Singular Value Decomposition (SVD) are particularly effective in reducing data complexity. It can represent high-dimensional data as a series of ordered feature vectors, thereby extracting the most representative information and reducing noise interference in air quality monitoring. This method has been widely used in the fields of image processing and signal analysis and has been proven to effectively reduce the dimensionality of complex environmental data and extract key features related to pollutant concentrations and changing trends (Zabalza et al. 2015).

Multidimensional decomposition techniques such as Tucker decomposition and tensor decomposition have further improved the feature extraction capabilities in air

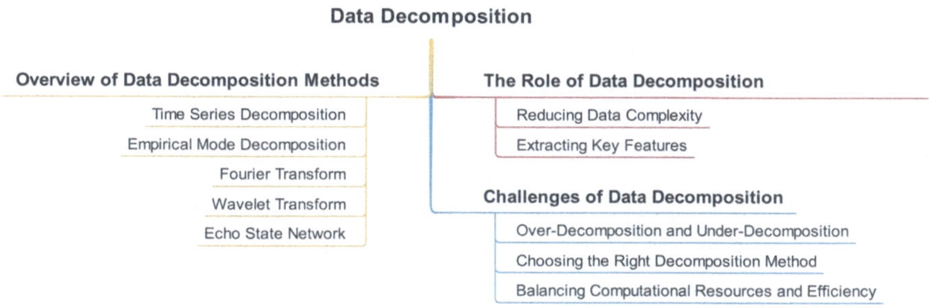

Fig. 1.8 Key issues and technologies in air quality data decomposition

quality monitoring. Since air quality data usually has multidimensional properties, such as changes in time, location, and pollutant type, the use of tensor decomposition methods can simultaneously consider the relationship between multiple dimensions, thereby retaining more data structure information. In addition to traditional linear decomposition methods, nonlinear decomposition methods such as Variational Mode Decomposition (VMD) and Empirical Mode Decomposition (EMD) show great potential in processing nonlinear and non-stationary signals. VMD can effectively extract useful features in nonlinear systems by decomposing complex nonlinear signals into multiple intrinsic mode functions. Similarly, methods such as EMD and Local Mean Decomposition (LMD) can effectively eliminate noise and extract useful features by recursively decomposing the intrinsic modes of signals.

When processing high-dimensional and noisy data, sparse decomposition technology provides an effective dimensionality reduction and feature extraction method. Sparse Singular Value Decomposition (SSVD) can effectively compress redundant information in the data and extract the most important features by sparsely processing the data. In addition, matrix rank reduction techniques based on SVD and Principal Component Analysis (PCA) are also widely used in dimensionality reduction and feature extraction of text and image data, which can significantly reduce data complexity and improve the performance of classification and recognition tasks.

In general, the application of data decomposition methods in air quality monitoring not only improves the accuracy of feature extraction but also greatly reduces the complexity of data. By decomposing high-dimensional, nonlinear, and non-stationary data into easy-to-process low-dimensional features, data decomposition technology provides strong technical support for real-time monitoring and emergency response. In air quality monitoring systems, decomposition methods can not only extract meaningful features under high noise backgrounds but also improve monitoring efficiency and response speed by reducing data dimensions.

1.2.2.2 Challenges of Data Decomposition

In air quality monitoring, data decomposition is the core link of feature extraction, however, choosing the right decomposition method has always been a complex and challenging problem. Air quality data usually contains multiple complex signal sources, including long-term trends, seasonal changes, and short-term pollution peaks. Different decomposition techniques perform differently in dealing with these complexities, so choosing the right method is crucial. However, the problems of over-decomposition and under-decomposition often affect the accuracy of the decomposition results. When over-decomposition occurs, the data is over-segmented, causing important signal features to be submerged in noise, while under-decomposition may prevent important information from being separated, resulting in the model being unable to accurately capture the true dynamics of environmental variables. This challenge is prevalent in various decomposition methods, whether it is SVD or VMD, it is necessary to carefully balance the depth of decomposition and the effectiveness of information extraction.

In the practical application of data decomposition, it is difficult to make tradeoffs among different decomposition methods to meet the needs of specific tasks. Taking air quality monitoring as an example, the nonlinearity and non-stationarity of the signal require the decomposition method to effectively extract dynamic change features, which usually requires a combination of multiple decomposition techniques. In speech signal feature extraction, Tucker decomposition has been shown to have strong feature extraction capabilities in multi-dimensional signal processing, but its complexity reduces processing efficiency when the amount of data is large (Yang et al. 2013). Similarly, variational mode decomposition performs well in extracting nonlinear signal features, but the choice of the number of decomposition layers has a great impact on the stability of the results. Too many decomposition layers may lead to data redundancy, while insufficient decomposition layers may miss key features (Zamora et al. 2019). Therefore, in practical applications, the choice of decomposition method needs to balance the complexity of the data, the limitation of computing resources, and the efficiency of feature extraction.

1.2.2.3 Overview of Data Decomposition Methods

Data decomposition methods are increasingly used in environmental monitoring, especially in the processing of air pollution data. These methods can not only effectively capture trends and periodicity in the data, but also separate short-term fluctuations and noise in complex data, thus providing a clearer structure for the analysis and prediction of environmental data. Table 1.2 shows the principles and applicable scenarios of different methods such as time series decomposition, empirical mode decomposition, and Fourier transform.

Time series decomposition methods help researchers identify long-term pollution trends and periodic changes by dividing data into trend, seasonal, and residual parts, especially when analyzing the concentration changes of pollutants such as PM2.5, it can reveal the regularity behind them. In addition, EMD is an adaptive decomposition method for processing nonlinear and non-stationary signals, which plays an important role in air quality prediction. By decomposing complex pollution data into different modes, EMD can effectively capture the local fluctuation characteristics in the data, thereby improving the accuracy of prediction. EMD is particularly suitable for analyzing short-term fluctuations in pollutant concentrations, helping to better understand the suddenness of pollution events (Wu and Lin 2019; Jiang et al. 2022).

The Fourier transform helps researchers analyze the periodic components of air pollutants by converting time domain signals into frequency domain signals. This method is particularly suitable for studying the laws of pollutant generation and dissipation, and through frequency analysis, it reveals the long-term fluctuation patterns of pollutants in the atmosphere (Salcedo et al. 1999; Sebald et al. 2000).

Table 1.2 Overview of data decomposition methods

Decomposition method	Principle	Applicable scenarios	Literature support
Time series decomposition	Decomposes data into trend, seasonal, and residual components, used for long-term trend analysis and forecasting	Analysis of trends in air pollutant concentration	WEST (1997); Salcedo et al. (1999)
EMD	Decomposes signals into multiple intrinsic mode functions, suitable for nonlinear, non-stationary data	PM2.5 concentration prediction and analysis	Wu and Lin (2019); Jiang et al. (2022)
Fourier transform	Converts time-domain signals into frequency-domain signals, useful for analyzing periodic changes	Periodic analysis of pollutant generation and dispersion	Salcedo et al. (1999); Sebald et al. (2000)
Wavelet transform	Decomposes data into different frequency components while retaining time information, suitable for analyzing both short- and long-term changes	Analysis of sudden events and trend changes in air pollution	Feng et al. (2015)
ESN	A time series forecasting model based on random recurrent neural networks, combined with decomposition methods	Multi-step PM2.5 concentration forecasting	Liu et al. (2020a)
Empirical wavelet transform (EWT)	Combines wavelet and EMD to decompose different frequency and time domain features	Air pollution forecasting	Liu et al. (2020b)

The wavelet transform provides great convenience for capturing short-term and long-term changes in air pollution data. Unlike the Fourier transform, the wavelet transform can not only perform frequency analysis but also retain the temporal information of the data, which makes it particularly effective in dealing with sudden pollution events. The wavelet transform can identify rapid changes and long-term trends in pollutant concentrations through multi-scale analysis, thereby providing more levels of information support for prediction. In addition, the Echo State Network (ESN) combines the advantages of recurrent neural networks and can be used in conjunction with decomposition methods to process multi-step predictions of air pollution. By decomposing air pollution data and making predictions, ESN has shown significant improvements in prediction accuracy (Liu et al. 2020a).

The EWT combines the advantages of wavelet transform and EMD and can decompose and analyze data in both the frequency domain and the time domain, which is particularly suitable for multi-step predictions of air pollution. By decomposing pollution data into multiple frequency components, EWT can more accurately extract the changing pattern of pollutant concentrations, thereby improving the accuracy of predictions (Liu et al. 2020b).

Through these data decomposition methods, air pollution predictions have not only become more accurate but also can deeply reveal the potential mechanisms of pollutant changes. The combined use of these methods enables researchers to analyze and process air pollution data from multiple dimensions, providing an important scientific basis for environmental management and policymaking.

1.2.3 Data Identification

Data recognition plays a vital role in the analysis and decision-making process of air quality monitoring. Figure 1.9 shows the core functions of data recognition in locating pollution sources and identifying pollution patterns, highlights the challenges faced when applying these technologies, and outlines its important role in tracking pollution diffusion paths, optimizing monitoring networks, and revealing cyclical changes in pollution.

1.2.3.1 Importance of Data Identification in Air Quality Monitoring

Data identification refers to the discovery and analysis of hidden structures and patterns from complex, multidimensional data sets through algorithms or models. It conducts in-depth analysis and interpretation of air pollution data by applying machine learning, statistical modeling, and data mining techniques within the framework of data science. The application of data identification in air quality monitoring covers many aspects, from locating pollution sources to identifying pollutant patterns to providing a scientific basis for environmental policymaking.

Data identification in air quality monitoring is crucial for locating pollution sources. In a monitoring network, data identification technology can help locate pollution sources in a specific area and track the diffusion path of pollutants. By analyzing the changes in the concentration of specific pollutants in the air, data identification technology can match pollutants with their sources and evaluate the

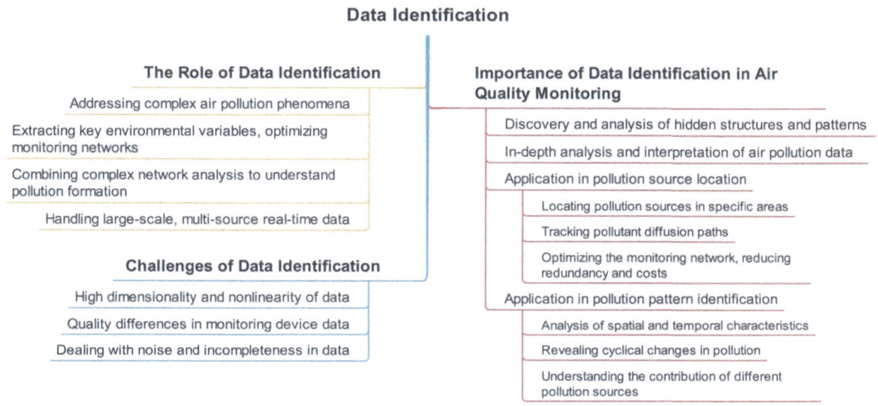

Fig. 1.9 Key issues and technologies in air quality data identification

contribution of emission sources. This not only helps to understand the source of pollutants but also detects redundant or unnecessary monitoring sites, thereby optimizing the monitoring network and reducing costs. At the same time, data identification technology also plays an important role in pollution pattern identification. Air pollution is usually affected by a combination of multiple factors, including meteorological conditions, geographical location, and emission patterns of pollution sources. Data identification helps reveal the spatial and temporal characteristics of pollution patterns through multidimensional data analysis. By identifying these patterns, researchers can better understand the cyclical changes in pollution and the contribution of different pollution sources to the overall pollution level.

Data identification also plays a vital role in air quality management and policymaking. Air quality monitoring data provides policymakers with a reliable scientific basis to help them formulate and implement effective pollution control measures. By identifying and analyzing historical data, policymakers can evaluate the impact of different policies on pollution levels and formulate adjustments based on new data.

In general, data identification plays a central role in air quality monitoring. It not only helps researchers understand the complex dynamics of pollutants, but also provides strong support for pollution source identification, in-depth analysis of pollution patterns, and policymaking. With the continuous advancement of data science and technology, data identification methods will further enhance our understanding of air pollution, thereby promoting more effective air quality management and policy implementation.

1.2.3.2 The Role and Challenges of Data Identification

Data identification plays a vital role in air quality monitoring, helping us better understand and analyze complex air pollution phenomena. Air pollution is usually caused by multiple pollution sources and involves a large number of environmental variables and meteorological factors, which makes it difficult to directly identify pollution patterns through simple observations. Data identification technology can reveal the complex relationships behind these pollution phenomena by extracting key information from massive amounts of environmental data. For example, through methods such as principal component analysis and cluster analysis, redundant information in air monitoring sites can be effectively identified, the layout of the monitoring network can be optimized, unnecessary costs can be reduced, and the integrity and accuracy of pollution source data can be ensured. Such data processing methods not only help locate the main pollution sources but also improve the accuracy of air quality assessment, thereby providing strong data support for decision-making (Wang et al. 2018). In addition, data identification can also combine complex network analysis technology to extract key pollution factors from multiple monitoring point data, and by analyzing the relationship between these factors, better understand the formation mechanism and propagation path of air pollution (Fan et al. 2016).

The advantage of data identification is that it can process large-scale, multi-source real-time data, especially in the context of the widespread application of modern smart sensors and Internet of Things technologies. Smart sensor networks can collect high-resolution data of various pollutants including PM2.5 and nitrogen dioxide in real-time, and data identification technology can convert these data into actionable information through effective data processing and classification methods.

Despite the increasing application of data recognition technology in air quality monitoring, it still faces some key challenges. First, air quality data itself has characteristics such as high dimensionality and nonlinearity, which brings great complexity to data recognition. Due to the complexity of pollution sources and their interactions, traditional linear data processing methods have difficulty capturing these complex dynamic relationships. In addition, the difference in data quality between different monitoring devices is also a challenge for data recognition, especially when using low-cost sensors, the accuracy and stability of the sensors are often not guaranteed, which may lead to inconsistent data and noise, which in turn affects the identification and prediction of pollution patterns. Another important challenge is how to deal with noise and incompleteness in the data. Air quality data will inevitably be affected by external factors such as weather conditions and equipment failures during the collection process, which will cause missing or abnormal data, thus affecting the overall analysis results.

1.2.4 Data Clustering

Figure 1.10 shows the importance of data clustering in air quality monitoring, the main challenges it faces, and an overview of common data clustering methods. Cluster analysis can identify hidden patterns in air pollution data and reveal the spatiotemporal variation characteristics of pollution sources. At the same time, the challenges of data clustering are mainly reflected in the complexity, dynamics, high dimensionality, and nonlinearity of the data. There are various methods to overcome these challenges, including partition clustering, hierarchical clustering, density-based clustering, and time series clustering technology.

1.2.4.1 The Significance of Data Clustering in Air Quality Monitoring

Cluster analysis is an unsupervised machine learning method that can help identify hidden patterns and similarities in air pollution data. Through clustering algorithms, such as K-means and hierarchical clustering, researchers can classify monitoring sites or periods into different categories based on air quality index or pollutant concentration. This classification helps to discover the spatiotemporal variation characteristics of air quality and the impact of different pollution sources. Data clustering can be used to analyze regional pollution characteristics,

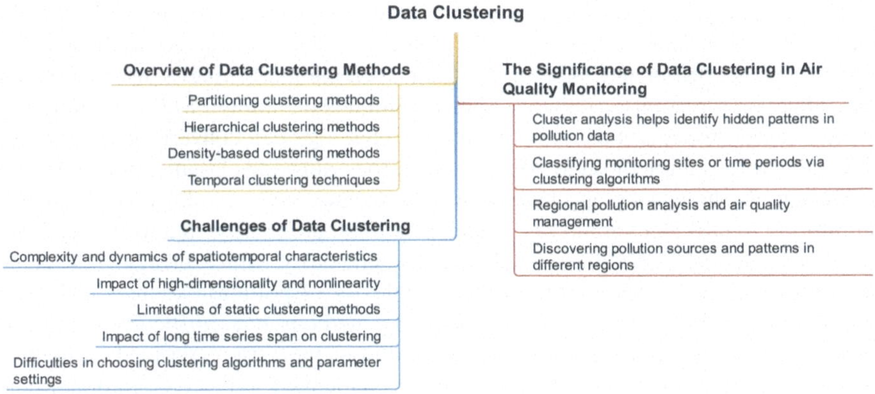

Fig. 1.10 Key issues and technologies in air quality data clustering

especially when studying large-scale regional pollution. Different regions have different pollution sources and meteorological conditions. Clustering technology can help analyze these differences and provide a basis for regional air quality management. By clustering data from multiple monitoring sites, areas with similar pollution sources can be identified and pollution patterns in different regions can be discovered. For example, Stolz et al. (2020) analyzed the air quality monitoring network of major cities in Mexico through clustering methods, successfully identified redundant sites, and revealed the pollution characteristics of each city (Stolz et al. 2020). Ignaccolo et al. (2008) also found similarities between air pollution in different monitoring sites in Piedmont, Italy through functional clustering analysis, revealing the spatial pattern of regional pollution characteristics (Ignaccolo et al. 2008).

The spatiotemporal correlation of air quality data is a key element in cluster analysis. Air pollution is affected by many factors, including seasonal changes, meteorological conditions, and human activities, resulting in complex correlations in air quality data in time and space. Cluster analysis can identify pollution characteristics at different time and space ranges. For example, Cheam et al. (2017) proposed a clustering model based on spatiotemporal data and applied it to air quality monitoring around Paris (Cheam et al. 2017). They found the variation patterns of nitrogen oxides at different periods of the day and revealed the spatiotemporal correlation of air pollution. Huang (1992) used stepwise cluster analysis to predict air quality in Xiamen, China and analyzed the characteristics of pollution changes at different times and regions (Huang 1992). These studies have shown that spatiotemporal cluster analysis can reveal the dynamic changes of pollutant concentrations over time and place, which helps to understand the spatiotemporal propagation patterns of air pollution.

1.2.4.2 Main Challenges of Data Clustering

The spatiotemporal characteristics bring significant challenges to data clustering, mainly reflected in the complexity and dynamics of the data. Air quality data not only have significant differences in space but also have different pollution sources and diffusion patterns in each region due to different geographical locations, topography, climate, and other factors; at the same time, the time dimension also brings frequent fluctuations in seasonality, weather changes, and human activities. This spatiotemporal interaction characteristic makes it difficult for traditional static clustering methods to cope with it. Under the spatiotemporal characteristics, air pollution data exhibits high dimensionality and nonlinear characteristics, which further increases the difficulty of data clustering. Traditional clustering methods, such as K-means or hierarchical clustering, usually assume that data has a linear relationship in a certain dimension, but for air quality data, the diffusion and interaction of pollutants often show nonlinear and complex coupling relationships. This complexity makes it impossible for simple distance measurement methods to accurately capture the similarities or differences between different sites.

The spatiotemporal characteristics have an important impact on the results of clustering analysis. Due to the obvious spatiotemporal dynamics of air pollution, traditional static clustering methods may not be able to capture these changes. Spatiotemporal clustering technology can be combined with time series analysis to accurately reveal the changes in pollutants in different periods and locations. Fan et al. (2018) analyzed Beijing's air quality data through a visibility graph network, revealing the fluctuation pattern of air quality in different periods, and further verified the impact of spatiotemporal characteristics on cluster analysis (Fan et al. 2018). These studies emphasize the importance of spatiotemporal characteristics in pollution data analysis. More accurate and meaningful clustering results can be obtained only by fully considering these factors.

The long time series data also poses challenges to data clustering. Over time, trends, seasonal fluctuations, and random disturbance factors in air quality data will cause changes in data structure. If the clustering algorithm does not fully consider the data characteristics of the time dimension, the clustering results may be distorted. During high pollution events, pollutant concentrations rise sharply but remain at a low level during regular periods. Such extreme events often affect the accuracy of clustering results.

There are often many difficulties in the selection and parameter setting of clustering algorithms. These difficulties mainly stem from the complexity, multidimensionality, and heterogeneity of air quality data. Different clustering algorithms perform differently when processing air quality data. For example, the K-means clustering algorithm is a classic algorithm that is usually used to process relatively simple and evenly distributed data sets. However, air quality data is often highly heterogeneous and has a complex distribution structure due to the influence of multiple factors such as weather, terrain, and traffic. For such non-spherical or irregularly distributed data sets, the limitations of the K-means algorithm are particularly obvious, which may lead to biased or distorted clustering results. The Ensemble

Empirical Mode Decomposition, Principal Component Analysis, and Least Squares (EPLS) algorithm proposed by Chen et al. (2017) improves the performance of the traditional K-means algorithm on air quality data by introducing a noise processing mechanism, especially showing good performance in dealing with high noise and outliers (Chen et al. 2017). However, although this type of improved algorithm can improve the accuracy of clustering results, its parameter setting becomes more complicated, and how to balance the flexibility and stability of the algorithm becomes a difficult point.

At the same time, most clustering algorithms rely on certain preset parameters, and the parameter selection of clustering algorithms has an important impact on the final results. For example, K-means requires the number of clusters to be set in advance, while density clustering algorithms (such as DBSCAN) require parameters such as distance threshold and minimum number of points to be set. The selection of these parameters is often crucial to the clustering results, but in actual air quality monitoring, it becomes very difficult to reasonably set these parameters due to the volatility of pollutant concentrations and the spatial imbalance of monitoring sites. In their study, the scalability and computational efficiency of the algorithm are also key issues in data clustering in air quality monitoring. Air quality monitoring usually involves a large amount of spatiotemporal data, which is not only large in volume but also has high-dimensional characteristics. Traditional clustering algorithms perform well when processing small-scale data, but when faced with large-scale air quality data, the computational complexity increases rapidly, resulting in a significant increase in the algorithm running time. Soares et al. (2020) used a hierarchical clustering method to optimize the design of air quality monitoring sites in their study, but due to the large amount of data, the computational efficiency of the algorithm became one of its main limiting factors (Soares et al. 2020).

1.2.4.3 Overview of Common Data Clustering Methods

Data clustering methods play a key role in air quality monitoring. They can help analyze large-scale, multi-dimensional monitoring data and reveal the distribution patterns of pollutants in time and space. According to different principles and application scenarios, clustering methods can be roughly divided into partitioning, hierarchical, and density-based clustering. The following is a brief overview of these methods, detailing their working principles and application scenarios in air quality monitoring.

Partitioning Clustering Methods

K-means is the most commonly used partitioning clustering algorithm, which divides the data set into k pre-set clusters, and each data point is assigned to the cluster center closest to it. The algorithm optimizes the position of the cluster center through repeated iterations so that the points within the cluster are as close as possible and the points between clusters are as far away as possible. The advantages of K-means are simplicity and fast calculation speed, which are suitable for most application scenarios. However, its main disadvantages are that it is sensitive to noise and outliers, and the number of clusters needs to be set in advance. In air quality monitoring, K-means is widely used to identify pollution patterns and pollution source areas (Chen et al. 2017).

Hierarchical Clustering Method

Hierarchical clustering gradually merges or splits data by constructing a cluster tree (tree structure) to form a multi-level cluster structure. Hierarchical clustering is divided into two types: bottom-up (agglomerative) and top-down (divisive). Unlike K-means, hierarchical clustering does not require a preset number of clusters and can automatically generate clustering results at different levels, making it suitable for scenarios that need to explore pollution patterns at different levels. This method can better handle the complex structure of air quality data, especially in the identification of pollution sources in different geographical regions (Soares et al. 2020). The disadvantage of hierarchical clustering is that the computational complexity is high, especially when dealing with large-scale spatiotemporal data, the computational cost will increase rapidly.

Density-Based Clustering Methods

DBSCAN is a density-based clustering method that forms clusters by identifying high-density areas and treating low-density areas as noise. Unlike K-means, DBSCAN does not require a preset number of clusters and is more robust to noise and outliers. It is particularly suitable for processing air quality data with uneven spatial distribution, such as identifying pollution hotspots in cities or irregularly distributed pollution sources (Shetty et al. 2024). However, the effectiveness of DBSCAN depends on the selection of parameters, such as distance threshold and minimum number of points, which may be difficult to adjust in complex air quality monitoring data.

Temporal Clustering Technology

Temporal clustering technology is used to analyze the changing pattern of air quality over time and help identify pollution patterns within a specific period. For air quality monitoring, temporal clustering can reveal the dynamic changes in pollutant concentrations between day and night, weekdays and weekends, or seasons. Time series clustering can group similar periods together, thereby revealing long-term pollution trends. Dynamic Time Warping (DTW) is a technique commonly used in time series clustering. It processes asynchronous time data by measuring the dynamic similarity between time series (Suris et al. 2022). In air quality monitoring, DTW is widely used to compare pollution data from different periods and identify similar pollution patterns.

Table 1.3 is a comparison of different clustering methods. Various data clustering methods have their advantages and disadvantages in air quality monitoring. Researchers need to choose appropriate algorithms based on actual needs and data characteristics. By combining spatial and temporal clustering techniques, the spatiotemporal characteristics of air pollution can be analyzed more comprehensively, and the accuracy of monitoring and prediction can be improved.

Table 1.3 Comparison of different clustering methods

Clustering method	Principle	Advantages	Disadvantages	Application scenarios
K-means	Divides data into K predefined clusters and iteratively adjusts to minimize intra-cluster distance	High computational efficiency, suitable for spherical data	Sensitive to noise, requires pre-setting the number of clusters	Air quality data with simple structure and uniform distribution
Hierarchical clustering	Creates a dendrogram by gradually merging or splitting data points to form clusters	No need to pre-set the number of clusters, suitable for complex data structures	High computational complexity, and poor performance with large datasets	Optimizing monitoring network layout, and analyzing complex air quality data
DBSCAN	Identifies clusters by detecting dense regions and recognizes noise and outliers	Does not require pre-setting the number of clusters, performs well with non-spherical data, strong noise resistance	Complex parameter settings, limited performance with high-dimensional data	Irregularly distributed air quality data, and pollution event detection
Spatial clustering techniques	Analyzes the geographic spatial distribution of data to discover regional pollution characteristics	Can identify spatial pollution sources, and reveal pollution diffusion patterns	Requires spatial data, computational complexity depends on dataset size	Identifying pollution sources, and analyzing regional pollution characteristics
Temporal clustering techniques	Focuses on time series data patterns, analyzing dynamic changes in pollutant concentration over time	Can detect temporal patterns, suitable for detecting and predicting extreme events	Requires appropriate time window selection, challenging to handle data with large time spans	Analyzing temporal patterns of air pollution, predicting extreme events

1.2.5 Data Forecasting

As shown in Fig. 1.11, data prediction plays an important role in today's public health protection, especially in air quality prediction, which has a profound impact on public health. By combining traditional statistical models with machine learning and deep learning methods, prediction technology has been significantly improved, making it possible to process complex spatiotemporal data. However, there are also many challenges in the prediction process, including data complexity, incompleteness, and variable selection. Different prediction methods have their advantages, from linear models to deep learning, which are suitable for different data characteristics and application scenarios. These prediction technologies provide an important decision-making basis for air pollution warnings, financial market fluctuations, and other fields.

1.2.5.1 Impact of Air Quality Prediction on Public Health

Air quality prediction plays an increasingly important role in public health protection and is essential for preventing health risks. Pollutants including PM2.5, PM10, ozone, and sulfur dioxide have long been shown to be closely associated with morbidity and mortality from various respiratory and cardiovascular diseases. When air pollutant concentrations reach dangerous levels, short-term exposure may lead to aggravation of health problems, especially for susceptible populations such as the elderly, children, and patients with chronic diseases. Therefore, being able to predict air quality in a timely and accurate manner, especially changes in pollutant concentrations, can help people better take measures in advance to avoid health risks. In recent years, prediction technology has been significantly improved, and the accuracy of prediction models has been improved through the application of machine learning and deep learning algorithms. For example, models based on Bidirectional Long Short-Term Memory (BiLSTM) neural networks can effectively

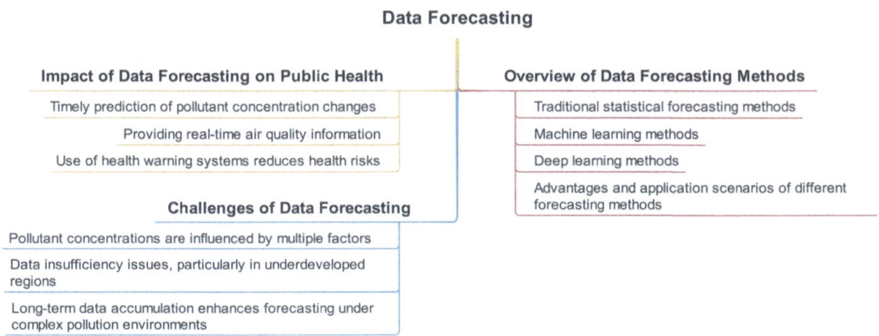

Fig. 1.11 Key issues and technologies in air quality data forecasting

predict changes in PM2.5 concentrations, which helps to identify high pollution periods in advance and reduce potential health threats (Zhang et al. 2021). In addition, machine learning methods such as random forests and support vector machines have also been used for air quality prediction. These models have improved the ability to predict changes in air quality under complex pollution environments through the accumulation of long-term data (Liang et al. 2020).

The results of air quality forecasts have played an important role in public health warnings and emergency responses. Many cities and regions provide real-time air quality information to the public through tools such as the Air Quality Index (AQI) to help people adjust their daily activities according to air conditions. Accurate forecasts can help local governments and public health agencies better plan emergency response measures, issue health warnings before high pollution events occur, and reduce the negative impact of air pollution on the public. For example, Canada's Air Quality Health Index (AQHI) has been widely used to help susceptible people reduce exposure risks. This system provides real-time health advice to the public through the comprehensive analysis of major air pollutants such as ozone, particulate matter, and nitrogen oxides (Gutenberg 2014). Similarly, in the United States, air quality forecasts are conducted through the National Air Quality Forecast Capability (NAQFC) system, which combines multiple forecast models to provide accurate warnings to areas with severe air pollution so that local governments can take measures to protect public health (Delle Monache et al. 2020). In addition, the forecast results are also used to formulate temporary measures such as traffic control and factory shutdowns to reduce emissions from pollution sources. By predicting and issuing health warnings in advance, the public can better plan outdoor activities and reduce exposure to highly polluted air, thereby reducing health risks associated with air pollution (Wu and Lin 2019). The application of this forecast-based warning system in public emergency response has not only improved overall social resilience but also significantly reduced the incidence of health problems caused by air pollution (Schurholz et al. 2020).

1.2.5.2 Challenges in Data Prediction

The main challenge in air quality prediction is the complexity of data, which places extremely high demands on prediction models. Air pollution data are often multidimensional and spatiotemporally heterogeneous, involving not only changes in the concentration of particulate matter (such as PM2.5, PM10) and gaseous pollutants (such as O_3, NO_x, SO_x, etc.), but also multiple variables such as meteorological data, geographical factors, and human activity patterns. The multi-source and dynamic nature of air quality data requires models to be able to effectively deal with nonlinear relationships between spatiotemporal data when processing. Time series data of air pollution are usually affected by multiple factors such as climate, transportation, and industrial emissions, resulting in high volatility and complexity in changes in pollutant concentrations. For example, Carmichael et al. (2008) pointed out that the complex interaction between pollutant emissions and atmospheric diffusion increases the difficulty of prediction. This complexity requires models to be able to integrate multiple observational data and deal with highly nonlinear pollution processes at the same time (Carmichael et al. 2008).

Insufficient data is another major challenge in air quality prediction, especially in underdeveloped areas or when high temporal resolution is required, where observational data are often incomplete or discontinuous. This lack of data makes it difficult to establish the model, especially in long-term trend forecasts and short-term extreme event forecasts, where the lack of data will increase the uncertainty of the model results. In addition, variable selection is another key issue in building air quality prediction models. When dealing with high-dimensional data, how to select the most representative and predictive variables is crucial to the performance of the model. The correlation between different variables and their contribution to pollutant concentrations is not always obvious, especially in the case of multi-source data. Selecting inappropriate variables may lead to excessive model complexity and increase computational costs, and may also cause overfitting problems, making the model fit the training data too accurately, but performing poorly when faced with new data.

1.2.5.3 Overview of Data Forecasting Methods

In the field of time series forecasting, traditional statistical models still play an important role in processing linear data and data with periodicity and trends. Common traditional methods such as the Autoregressive Integrated Moving Average (ARIMA) model can effectively process non-stationary time series data by combining autoregression and moving average methods. ARIMA transforms non-stationary data into stationary series through difference operations and then predicts future trends. This model performs well in areas such as air quality forecasting and can capture long-term trends and seasonal fluctuations in data. However, the performance of the ARIMA model is often limited when faced with highly nonlinear and complex time series. In contrast, the Generalized Autoregressive Conditional Heteroskedasticity (GARCH) model focuses on volatility forecasting, especially for financial market data. By modeling conditional heteroskedasticity in time series, GARCH can capture the characteristics of volatility over time and is therefore widely used in dealing with pollutant concentration forecasts with significant volatility. In addition, the Holt-Winters exponential smoothing model is good at processing time series with trends and seasonality. By separating the trend, seasonal components, and random error components, Holt-Winters can make accurate short-term and medium-term predictions and has been widely used in fields such as air quality management and energy consumption forecasting. Other methods such as the Vector Autoregression (VAR) model multiple mutually influencing variables and are often used in economic and financial fields that need to capture complex relationships between variables.

With the development of machine learning technology, the processing ability of nonlinear time series data has been significantly improved. Support Vector

Regression (SVR), as a regression model based on support vector machines, performs well in the prediction of time series with rich nonlinear features. SVR can effectively process complex nonlinear data by finding the optimal hyperplane in high-dimensional space, so it has been widely used in the prediction of air pollutant concentrations. At the same time, the random forest model, with the advantages of its ensemble learning method, reduces the overfitting problem by combining the prediction results of multiple decision trees and can handle large-scale and complex time series data. Compared with traditional linear models, random forests have a strong ability to capture nonlinear features, so they perform well in fields such as air quality and climate forecasting.

Deep learning models have shown strong capabilities in processing complex time series data, especially in capturing long-term dependencies and nonlinear relationships. LSTM networks are an improved version of Recurrent Neural Networks (RNNs) that are specifically designed to process long-term dependent time series data. By introducing memory units and gating mechanisms, LSTM networks can effectively capture long-term and short-term dependencies in sequences and are particularly suitable for applications such as air pollution prediction and financial data analysis with seasonal and trend characteristics. Compared with traditional methods and shallow machine learning models, LSTM networks can better handle complex.

Nonlinear data, especially in non-stationary time series. Although Convolutional Neural Networks (CNNs) are mainly used for image processing, they also show strong feature extraction capabilities in time series analysis. When used in combination with LSTM, CNN can extract spatiotemporal features in time series, while LSTM captures long-term dependencies. The combination of the two performs well in areas such as air quality prediction. The advantage of hybrid models is that they can take into account both local features and long-term trends in time series, which enables them to have higher prediction accuracy when dealing with highly complex data with obvious nonlinear characteristics.

Table 1.4 shows the basic principles of different forecasting methods and their typical application areas, indicating the evolution from traditional linear models to deep learning models, as well as the significant advantages of deep learning models when processing nonlinear time series data. The combined application of these methods makes time series forecasting models more flexible when facing complex data, and provides accurate decision support for fields such as air pollution and energy management.

1.2.6 Data Interpolation

Data interpolation technology plays a vital role in filling the gaps in spatiotemporal data, estimating pollution concentrations in unsampled areas, and improving monitoring coverage. The interpolation process also faces problems with data quality, spatial correlation, boundary areas, and outliers. Common interpolation methods

Table 1.4 Overview of data forecasting methods

Forecasting methods	Principle	Advantages	Disadvantages	Application scenarios
ARIMA	Combines AR and moving average models	Well-suited for linear data and stationary series	Struggles with nonlinear and highly volatile data	Air quality, financial forecasting
GARCH	Handle volatility and uncertainty in time series	Effectively models volatility and risk	Limited in handling non-volatility-related patterns	Financial markets, air quality forecasting
Holt-winters	Uses exponential smoothing	Excellent for seasonal data with trends	Assumes a constant seasonal pattern, limited in complex datasets	Air quality, water resource management
VAR	Captures interactions between multiple time series	Models interdependencies between variables	Requires a large number of parameters, computationally intensive	Economics, financial data
SVR	Fits a hyperplane in high-dimensional space to handle nonlinear time series	Strong performance with nonlinear relationships	Sensitive to choice of kernel, requires tuning	Air quality, economic forecasting
Random Forest	Reduce overfitting and improve generalization	Robust against overfitting, can handle nonlinearity	Can be slow with large datasets, and complex to interpret	Large-scale complex datasets
LSTM	Capture long-term dependencies in time series data	Captures long-term trends and dependencies in sequences	Computationally expensive, and requires a large amount of data	Air pollution forecasting, financial forecasting
CNN + LSTM hybrid model	Extracts spatial features and captures long-term dependencies in sequential data	Combines strengths of CNN and LSTM for high accuracy	Complex to implement and train, computationally intensive	Air quality, environmental monitoring

include Kriging interpolation, inverse distance weighted (IDW), and spline interpolation, and different methods are suitable for different data characteristics. Figure 1.12 shows the role of data interpolation in air quality monitoring, the main challenges faced, and common interpolation methods. The following will further explore the specific applications and technical details of data interpolation in air quality monitoring.

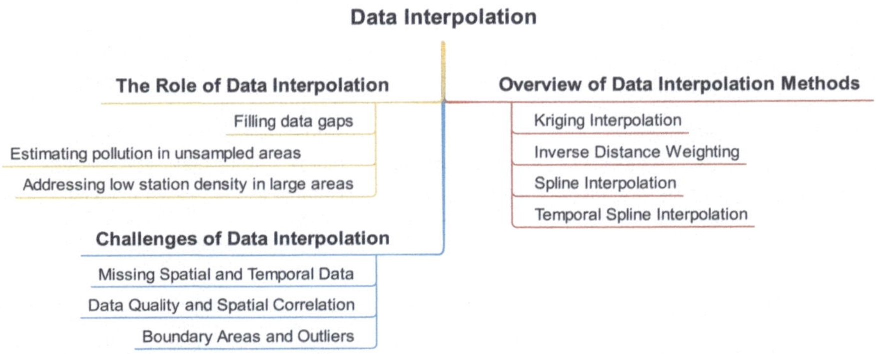

Fig. 1.12 Key issues and technologies in air quality data interpolation

1.2.6.1 The Role of Data Interpolation in Air Quality Monitoring

The role of data interpolation in air quality monitoring is crucial, especially in filling data gaps and providing continuous spatiotemporal data. Due to the limited distribution of air quality monitoring stations, monitoring data are usually discontinuous in space and time, which makes it difficult to assess the complete air quality. Interpolation techniques can estimate unsampled areas through existing monitoring data, provide more continuous spatiotemporal data, and make it easier to capture trends and changes in air pollution. Spatial interpolation techniques such as Kriging and IDW have been widely used to estimate pollution concentrations in unsampled areas. These methods significantly improve the spatiotemporal coverage and continuity of data by using known data from neighboring monitoring points to predict missing areas.

Data interpolation is also widely used to solve the problem of low density of monitoring stations, especially in air quality monitoring in larger areas. When monitoring stations are sparse, interpolation techniques can make up for the lack of spatial data by using data from surrounding stations. Taking residual Kriging as an example, this method not only uses existing data from monitoring stations but also provides more accurate estimates of spatial pollution concentrations by combining low-cost sensors and model prediction data. Such methods can effectively reduce the interpolation errors caused by the uneven distribution of monitoring stations, ensuring that the predicted data in unsampled areas are also highly accurate.

Interpolation results also play an important role in drawing air pollution distribution maps. Through spatial interpolation technology, researchers can convert limited monitoring data into pollution distribution maps for the entire region, which is particularly important in the visualization and trend analysis of air pollution. Pollution distribution maps not only provide intuitive references for decision-makers but also help identify pollution hotspots and provide a basis for the formulation of pollution control strategies. At present, pollution distribution maps generated based on Kriging have been widely used to assess air quality levels in different regions, such as in the spatial distribution analysis of atmospheric particulate matter (such as PM2.5, and PM10) and ozone concentrations. Such graphical results enable the public and decision-makers to clearly see the spatial distribution of pollution and take quick countermeasures when pollution incidents occur. By optimizing the interpolation algorithm, researchers can improve the accuracy of pollution distribution maps and further reduce spatial errors caused by the uneven distribution of monitoring stations (Van Egmond and Onderdelinden 1981; Candiani et al. 2013).

Through the application of these interpolation techniques, air quality monitoring can not only effectively fill in data gaps, but also draw more accurate pollution distribution maps, providing more comprehensive environmental data support for decision-makers. The continuous improvement and application of these technologies make air quality monitoring and governance more efficient and scientific.

1.2.6.2 Main Challenges of Data Interpolation

In air quality monitoring, data interpolation faces several key challenges, the first of which is the problem of missing spatial and temporal data. Due to the uneven distribution of monitoring sites, especially in large-scale monitoring networks, there are significant differences in monitoring coverage in different regions. The limited number of monitoring sites, especially in remote or geographically complex areas, often leads to a large number of data gaps. In addition, equipment failures and data transmission problems will further aggravate the data gap. This temporal and spatial imbalance makes it difficult for interpolation models to fill in data, especially when they need to rely on limited data provided by existing monitoring sites.

In addition to the problem of missing data, data quality, and spatial correlation also directly affect the accuracy of interpolation. In air quality monitoring, unstable data quality often stems from performance differences in monitoring site equipment, sensor calibration problems, and noise in the data collection process. Interpolation methods rely on high-quality input data, and when the data quality is poor, the performance of the interpolation model will be significantly reduced. At the same time, spatial correlation plays a vital role in the interpolation process. The low spatial correlation will result in the inability of data from neighboring areas to complement each other effectively, thereby reducing the interpolation accuracy. For example, Junninen et al. (2004) discussed the performance of different interpolation techniques in the face of missing data when evaluating air quality data interpolation methods, and emphasized the importance of spatial correlation in improving interpolation accuracy (Junninen et al. 2004).

Interpolation of boundary areas and outliers is also a prominent problem. In air quality monitoring, the lack of sufficient monitoring points in boundary areas makes it difficult for interpolation models to accurately capture the changes in pollutant concentrations in these areas. The existence of outliers, such as sudden pollution events or equipment error data, will cause the interpolation model to deviate from the actual situation and make it difficult to provide reliable predictions. Boundary areas are often at the edge of the monitoring network and lack sufficient spatial reference points, which increases the prediction instability of the interpolation model in the boundary area and is prone to large errors. When studying the relationship between monitoring network density and interpolation accuracy, Van Egmond and Onderdelinden (1981) pointed out that due to the lack of data support in boundary areas, interpolation errors are often large and cannot be solved by simple model adjustments (Van Egmond and Onderdelinden 1981). Janssen et al. (2008) also found that although the accuracy of some areas can be improved, there are still significant difficulties in dealing with boundary areas and outliers. When the abnormal data is inconsistent with the background trend, the interpolation results will deviate greatly from the actual situation (Janssen et al. 2008).

In general, the main challenges in air quality data interpolation include the lack of spatiotemporal data, the impact of data quality and spatial correlation on interpolation accuracy, and the difficulty of interpolation in boundary areas and outliers. These problems directly affect the effectiveness of the interpolation model and need to be solved through more complex algorithms and technical improvements.

1.2.6.3 Overview of Data Interpolation Methods

In environmental data processing, interpolation methods are essential tools for filling data gaps in space and time, helping scientists and researchers to infer values in unknown areas or time points. Spatial interpolation methods are mainly used to estimate data of unobserved points between known locations, while temporal interpolation infers missing values in time series.

Among the commonly used spatial interpolation methods, Kriging interpolation, IDW, and spline interpolation are the three most widely used methods. They are based on different theories and assumptions and are suitable for different data characteristics. Table 1.5 shows the characteristics and applicable scenarios of each method, which helps to choose the appropriate interpolation method to solve different problems.

Kriging interpolation is an interpolation method based on geo-statistics, assuming that there is a certain spatial correlation between variables. The spatial autocorrelation is analyzed by the semi-variogram, which can provide both predicted values and estimates of prediction errors. Kriging methods can be divided into Ordinary Kriging (OK), Universal Kriging (UK), and Kriging based Sequence Interpolation (KSI). Kriging interpolation has high accuracy when processing environmental data with strong spatial autocorrelation because it considers spatial correlation. It is especially suitable for estimating the spatial distribution of natural phenomena such as precipitation and temperature (Naoum and Tsanis 2004).

IDW is an interpolation method based on distance weighting. Its basic assumption is that the closer the point is, the greater the influence on the predicted point. IDW estimates the value of unknown points by using the inverse of the distance as the weight. It is suitable for processing discretely distributed data and is easy to implement. However, IDW does not consider spatial autocorrelation. For complex terrain or drastically changing phenomena, the effect is not as accurate as Kriging,

Table 1.5 Overview of data interpolation methods

Interpolation method	Principle	Advantages	Disadvantages	Application scenarios
Kriging interpolation	Based on the variogram, consider spatial correlation	Provides estimates and prediction error	Computationally complex, and time-consuming	Suitable for spatial distribution of pollutants with strong spatial correlation
Inverse distance weighting	Weighted by distance, closer points have more influence	Simple to compute, good for evenly distributed data	Does not account for spatial correlation, less precise	Suitable for evenly distributed air quality data
Ordinary kriging	Interpolates based on spatial correlation, and uses variance between known points	Provides spatial predictions and error estimates	Computationally intensive and requires detailed data	Ideal for high-precision pollutant distribution
Universal kriging	Similar to OK but accounts for external trends	Can incorporate external trends, adapt to complex environments	Complex model setup, longer computation times	Suitable for air pollutant concentrations affected by external factors
Linear interpolation	Draws a straight line between known points	Simple and fast, ideal for stable trends	Large error in non-linear data	Suitable for filling in missing values in stable series data
Lagrange interpolation	Uses higher-order polynomials to fit between data points	High accuracy for complex trends	Sensitive to noise, computationally intensive for large data sets	Suitable for detailed local pollutant concentration interpolation

but it performs well in applications such as climate data and precipitation data (Ikechukwu et al. 2017).

Spline interpolation estimates the value of unknown points by fitting a smooth curve (usually a quadratic or cubic polynomial) between known points. Its advantage is that it can generate continuous and smooth curves or surfaces, which is suitable for processing phenomena with smooth transitions, such as terrain data. However, spline interpolation may produce overfitting at boundaries or in areas where data changes drastically, affecting accuracy. Experiments show that spline interpolation has high accuracy in terrain interpolation, but performs poorly in data with weak spatial autocorrelation (Darmawan et al. 2023).

Linear interpolation is a simple and commonly used time interpolation method that estimates the value of an unknown point by drawing a straight line between two adjacent known points. Its advantages are simplicity and speed, and it is suitable for processing data with relatively stable trends, but it may cause large errors for data with nonlinear changes. Spline interpolation can also be used for time series data to estimate missing values by fitting a smooth polynomial function between adjacent points. Spline interpolation is suitable for processing time series data with smooth transitions, such as temperature and humidity, but it may cause overfitting problems in time series with drastic changes.

1.3 Scope of the Book

This book systematically introduces the application of data science in air quality monitoring, focusing on how to improve the accuracy of monitoring data and the effectiveness of decision-making through modern data processing and analysis techniques. From theory to application, this book covers all aspects of air quality monitoring, including data acquisition, preprocessing, decomposition, identification, clustering, prediction, and interpolation techniques. The book not only presents the theoretical basis of various data processing methods but also analyzes their performance in actual air quality monitoring through specific cases, striving to provide researchers and practitioners with comprehensive technical support.

The book is divided into seven chapters, each of which discusses a topic independently. At the same time, the contents of each chapter are interconnected to build a complete air quality monitoring data processing framework. The content of each chapter is organized as follows:

Chapter 1: Introduction

Chapter 1 lays the foundation for this book and outlines the basic concepts and importance of data science in air quality monitoring. This chapter first introduces the background and current status of air quality monitoring, emphasizing the core role of data in this field. Subsequently, key data processing issues in air quality monitoring are analyzed, including data collection, storage, and analysis. Finally, this chapter also briefly introduces the core technologies involved in subsequent chapters of this book, such as data decomposition, clustering, and prediction, and clarifies the scope and objectives of the book.

Chapter 2: Data Preprocessing in Air Quality Monitoring

Chapter 2 discusses in detail the importance and specific techniques of data preprocessing in air quality monitoring. This chapter first introduces the channels for obtaining air quality data and analyzes the characteristics of these data, especially the problems of missing data and outliers. Next, several commonly used missing value filling techniques, such as mean filling and interpolation, are introduced, and the advantages, disadvantages, and applicability of different methods are analyzed. This chapter also discusses the techniques of data anomaly detection and demonstrates the key role of these preprocessing techniques in improving data quality through performance comparison.

Chapter 3: Data Decomposition in Air Quality Monitoring

Chapter 3 focuses on data decomposition techniques and their application in air quality data analysis. This chapter first explains the concept and importance of data decomposition and then introduces wavelet decomposition and modal decomposition techniques in detail. Through examples, this chapter shows how these decomposition techniques can help scientists separate different signal components in air quality data, to conduct subsequent analysis more accurately. In addition, this chapter also compares the performance of different decomposition methods through experiments to help readers choose the most suitable decomposition strategy.

Chapter 4: Data Identification in Air Quality Monitoring

Chapter 4 discusses the application of data identification technology in air quality monitoring, especially the latest progress in feature selection and feature extraction. This chapter first introduces how to use feature selection technology to screen out variables that have a significant impact on air quality, thereby improving the efficiency and accuracy of the model. Next, different methods of feature extraction, such as PCA and Linear Discriminant Analysis (LDA), are analyzed and their application effects are demonstrated through experiments. The last part of this chapter combines these techniques with practical application scenarios to help readers understand how to use these techniques for data identification in actual monitoring.

Chapter 5: Data Clustering in Air Quality Monitoring

Chapter 5 focuses on data clustering technology and its application in air quality monitoring, especially the implementation methods of temporal clustering and spatial clustering. This chapter first introduces the basic concept of data clustering and then explores how temporal clustering can help researchers identify the laws of air quality changes in different periods, while spatial clustering is used to analyze the distribution patterns of air pollution between different geographical regions. Through case analysis, this chapter demonstrates the application effects of various clustering algorithms and evaluates the advantages and disadvantages of these methods through performance comparison.

Chapter 6: Data Prediction in Air Quality Monitoring

Chapter 6 explores air quality data prediction technology in depth and analyzes how to effectively predict future air quality trends from the perspectives of deterministic prediction and probabilistic prediction. This chapter first introduces common deterministic prediction models, such as linear regression and time series analysis, and then introduces probabilistic prediction models, especially prediction methods based on machine learning such as random forests and neural networks. Through performance comparison, this chapter analyzes the accuracy and computational efficiency of different models and combines actual cases to demonstrate the application of these prediction methods in air quality prediction.

Chapter 7: Data Interpolation in Air Quality Monitoring

Chapter 7 discusses the importance of data interpolation technology in air quality monitoring, especially the application of time interpolation and spatial interpolation. This chapter first introduces time interpolation techniques, such as linear interpolation and spline interpolation, and analyzes their performance in dealing with missing values in time series. Next, this chapter focuses on spatial interpolation techniques, such as Kriging and IDW, and demonstrates their application in spatial data estimation. Through performance comparison, this chapter summarizes the applicability and effects of different interpolation methods and provides readers with specific suggestions for selecting interpolation techniques.

References

Ameer S, Shah MA, Khan A, Song H, Maple C, ul Islam S, Asghar MN (2019) Comparative analysis of machine learning techniques for predicting air quality in smart cities. IEEE Access 7:128325–128338. https://doi.org/10.1109/ACCESS.2019.2925082

Arroyo P, Herrero JL, Suarez JI, Lozano J (2019) Wireless sensor network combined with cloud computing for air quality monitoring. Sensors 19(3):691. https://doi.org/10.3390/s19030691

Bell ML, Morgenstern RD, Harrington W (2011) Quantifying the human health benefits of air pollution policies: review of recent studies and new directions in accountability research. Environ Sci Pol 14(4):357–368. https://doi.org/10.1016/j.envsci.2011.02.006

Benammar M, Abdaoui A, Ahmad SHM, Touati F, Kadri A (2018) A modular IoT platform for real-time indoor air quality monitoring. Sensors 18(2):581. https://doi.org/10.3390/s18020581

Blair GS, Henrys P, Leeson A, Watkins J, Eastoe E, Jarvis S, Young PJ (2019) Data science of the natural environment: A research roadmap. Front Environ Sci 7:121. https://doi.org/10.3389/fenvs.2019.00121

Brammer JR, Brunet ND, Burton AC, Cuerrier A, Danielsen F, Dewan K, Herrmann TM, Jackson MV, Kennett R, Larocque G, Mulrennan M, Pratihast AK, Saint-Arnaud M, Scott C, Humphries MM (2016) The role of digital data entry in participatory environmental monitoring. Conserv Biol 30(6):1277–1287. https://doi.org/10.1111/cobi.12727

Brantley HL, Hagler GSW, Kimbrough ES, Williams RW, Mukerjee S, Neas LM (2014) Mobile air monitoring data-processing strategies and effects on spatial air pollution trends. Atmos Meas Tech 7(7):2169–2183. https://doi.org/10.5194/amt-7-2169-2014

Burr W, Newlands NK, Zammit-Mangion A (2023) Environmental data science: part 2. Environmetrics 34(2):e2788. https://doi.org/10.1002/env.2788

Candiani G, Carnevale C, Finzi G, Pisoni E, Volta M (2013) A comparison of reanalysis techniques: applying optimal interpolation and ensemble Kalman filtering to improve air quality monitoring at mesoscale. Sci Total Environ 458:7–14. https://doi.org/10.1016/j.scitotenv.2013.03.089

Carmichael GR, Sandu A, Chai T, Daescu DN, Constantinescu EM, Tang Y (2008) Predicting air quality: improvements through advanced methods to integrate models and measurements. J Comput Phys 227(7):3540–3571. https://doi.org/10.1016/j.jcp.2007.02.024

Castell N, Dauge FR, Schneider P, Vogt M, Lerner U, Fishbain B, Broday D, Bartonova A (2017) Can commercial low-cost sensor platforms contribute to air quality monitoring and exposure estimates? Environ Int 99:293–302. https://doi.org/10.1016/j.envint.2016.12.007

Cheam ASM, Marbac M, McNicholas PD (2017) Model-based clustering for spatiotemporal data on air quality monitoring. Environmetrics 28(3):e2437. https://doi.org/10.1002/env.2437

Chen B, Kan H (2008) Air pollution and population health: a global challenge. Environ Health Prev Med 13(2):94–101. https://doi.org/10.1007/s12199-007-0018-5

Chen Y, Wang L, Li F, Du B, Choo KKR, Hassan H, Qin W (2017) Air quality data clustering using EPLS method. Information Fusion 36:225–232. https://doi.org/10.1016/j.inffus.2016.11.015

Choirat C, Braun D, Kioumourtzoglou MA (2019) Data science in environmental health research. Curr Epidemiol Rep 6(3):291–299. https://doi.org/10.1007/s40471-019-00205-5

Darmawan Y, Munawar ADA, Wahyujati H, Nainggolan L (2023) Accuracy assessment of spatial interpolations methods using ArcGIS. E3s Web of Conferences 464:09005. https://doi.org/10.1051/e3sconf/202346409005

Delle Monache L, Alessandrini S, Djalalova I, Wilczak J, Knievel JC, Kumar R (2020) Improving air quality predictions over the United States with an analog ensemble. Weather Forecast 35(5):2145–2162. https://doi.org/10.1175/WAF-D-19-0148.1

Evagelopoulos V, Charisiou ND, Begou P (2023) Fault detection of air quality measurements using artificial intelligence. E3s Web of Conferences 436:10005. https://doi.org/10.1051/e3sconf/202343610005

Fan X, Wang L, Xu H, Li S, Tian L (2016) Characterizing air quality data from complex network perspective. Environ Sci Pollut Res 23(4):3621–3631. https://doi.org/10.1007/s11356-015-5596-y

Fan X, Zhang Q, Wang L, Yin J (2018) Visibility graph network analysis of air quality data. Int J Environ Monit Anal 6(3):110–115. https://doi.org/10.11648/j.ijema.20180603.15

Feng X, Li Q, Zhu Y, Hou J, Jin L, Wang J (2015) Artificial neural networks forecasting of $PM_{2.5}$ pollution using air mass trajectory based geographic model and wavelet transformation. Atmos Environ 107:118–128. https://doi.org/10.1016/j.atmosenv.2015.02.030

Giudici P (2018) Financial data science. Statistics Probability Lett 136:160–164. https://doi.org/10.1016/j.spl.2018.02.024

Gutenberg S (2014) Demystifying the air quality health index. Can Pharm J 147:332–334. https://doi.org/10.1177/1715163514552560

Huang G (1992) A stepwise cluster analysis method for predicting air quality in an urban environment. Atmos Environ B Urb Atmos 26(3):349–357. https://doi.org/10.1016/0957-1272(92)90010-P

Ignaccolo R, Ghigo S, Giovenali E (2008) Analysis of air quality monitoring networks by functional clustering. Environmetrics 19(7):672–686. https://doi.org/10.1002/env.946

Ikechukwu MN, Ebinne E, Idorenyin U, Raphael NI (2017) Accuracy assessment and comparative analysis of IDW, spline and kriging in spatial interpolation of landform (topography): an experimental study. J Geogr Inf Syst 9(3):354–371. https://doi.org/10.4236/jgis.2017.93022

Janssen S, Dumont G, Fierens F, Mensink C (2008) Spatial interpolation of air pollution measurements using CORINE land cover data. Atmos Environ 42(20):4884–4903. https://doi.org/10.1016/j.atmosenv.2008.02.043

Jiang W, Zhu G, Shen Y, Xie Q, Ji M, Yu Y (2022) An empirical mode decomposition fuzzy forecast model for air quality. Entropy 24(12):1803. https://doi.org/10.3390/e24121803

Junninen H, Niska H, Tuppurainen K, Ruuskanen J, Kolehmainen M (2004) Methods for imputation of missing values in air quality data sets. Atmos Environ 38(18):2895–2907. https://doi.org/10.1016/j.atmosenv.2004.02.026

Krzyzanowski M, Vandenberg J, Stieb D (2005) Perspectives on air quality policy issues in Europe and North America. J Toxicol Environ Health A 68(13–14):1057–1061. https://doi.org/10.1080/15287390590935897

Kulikova E, Sulimin V, Shvedov V (2023) Artificial intelligence for ambient air quality control. E3s Web of Conferences 419:03011. https://doi.org/10.1051/e3sconf/202341903011

Kurt OK, Zhang J, Pinkerton KE (2016) Pulmonary health effects of air pollution. Curr Opin Pulm Med 22(2):138–143. https://doi.org/10.1097/MCP.0000000000000248

Liang YC, Maimury Y, Chen AHL, Juarez JRC (2020) Machine learning-based prediction of air quality. Appl Sci 10(24):9151. https://doi.org/10.3390/app10249151

Liu H, Long Z, Duan Z, Shi H (2020a) A new model using multiple feature clustering and neural networks for forecasting hourly PM2.5 concentrations, and its applications in China. Engineering 6(8):944–956. https://doi.org/10.1016/j.eng.2020.05.009

Liu H, Yin S, Chen C, Duan Z (2020b) Data multi-scale decomposition strategies for air pollution forecasting: A comprehensive review. J Clean Prod 277:124023. https://doi.org/10.1016/j.jclepro.2020.124023

Liu N, Liu X, Jayaratne R, Morawska L (2020c) A study on extending the use of air quality monitor data via deep learning techniques. J Clean Prod 274:122956. https://doi.org/10.1016/j.jclepro.2020.122956

Manfreda S, McCabe MF, Miller PE, Lucas R, Madrigal VP, Mallinis G, Dor EB, Helman D, Estes L, Ciraolo G, Mullerova J, Tauro F, Isabel de Lima M, de Lima JLMP, Maltese A, Frances F, Caylor K, Kohv M, Perks M, Ruiz-Perez G, Su Z, Vico G, Toth B (2018) On the use of unmanned aerial systems for environmental monitoring. Remote Sens 10(4):641. https://doi.org/10.3390/rs10040641

Martinez Bilesio AR, Batistelli M, Garcia-Reiriz AG (2019) Fusing data of different orders for environmental monitoring. Anal Chim Acta 1085:48–60. https://doi.org/10.1016/j.aca.2019.08.005

Miasayedava L, Kaugerand J, Tuhtan JA (2023) Lightweight open data assimilation of Pan-European urban air quality. IEEE Access 11:84670–84688. https://doi.org/10.1109/ACCESS.2023.3302348

Naoum S, Tsanis IK (2004) A hydroinformatic approach to assess interpolation techniques in high spatial and temporal resolution. Can Water Resour J 29(1):23–46. https://doi.org/10.4296/cwrj23

Owusu PA, Sarkodie SA (2020) Global estimation of mortality, disability-adjusted life years and welfare cost from exposure to ambient air pollution. Sci Total Environ 742:140636. https://doi.org/10.1016/j.scitotenv.2020.140636

Rasool M, Chaudhary VK (2022) Applications of data science in respective engineering domains. Int J Sci Res Sci Technol 9(5):71–75. https://doi.org/10.32628/IJSRST22958

Roosken AAM (1978) Real time validation of air quality data. In: Studies in environmental science, vol 2. Elsevier, pp 85–89. https://doi.org/10.1016/S0166-1116(08)70815-X

Salcedo RLR, Alvim Ferraz MCM, Alves CA, Martins FG (1999) Time-series analysis of air pollution data. Atmos Environ 33(15):2361–2372. https://doi.org/10.1016/S1352-2310(99)80001-6

Saracco BH (2020) Data science and predictive analytics: biomedical and health applications using R. J Med Libr Assoc 108(2):334–334. https://doi.org/10.5195/jmla.2020.901

Schurholz D, Kubler S, Zaslavsky A (2020) Artificial intelligence-enabled context-aware air quality prediction for smart cities. J Clean Prod 271:121941. https://doi.org/10.1016/j.jclepro.2020.121941

Sebald L, Treffeisen R, Reimer E, Hies T (2000) Spectral analysis of air pollutants. Part 2: ozone time series. Atmos Environ 34(21):3503–3509. https://doi.org/10.1016/S1352-2310(00)00147-3

Shetty C, Seema S, Sowmya BJ, Nandalike R, Supreeth S, Dayananda P, Rohith S, Vishwanath Y, Ranjan R, Goud V (2024) A machine learning approach for environmental assessment on air quality and mitigation strategy. J Eng 2024:2893021. https://doi.org/10.1155/2024/2893021

Shukla PK, Sharmila, Singh A, Shaikh N, Sharma A, Singh R (2023) Air pollution monitoring by indulging AI and IoT for environmental protection. In: 2023 3rd international conference on pervasive computing and social networking (ICPCSN), 6/2023. IEEE, Salem, pp 1161–1165. https://doi.org/10.1109/ICPCSN58827.2023.00197

Simo A, Dzitac S, Frigura-Iliasa FM, Musuroi S, Andea P, Meianu D (2020) Technical solution for a real-time air quality monitoring system. Int J Computer Commun Control 15(4):3891. https://doi.org/10.15837/ijccc.2020.4.3891

Soares J, Makar P, Aklilu YA, Akingunola A (2020) Hierarchical clustering for optimizing air quality monitoring networks. In: Mensink C, Gong W, Hakami A (eds) Air pollution modeling and its application XXVI. Springer International Publishing, Cham, pp 299–303. https://doi.org/10.1007/978-3-030-22055-6_47

Stolz T, Huertas ME, Mendoza A (2020) Assessment of air quality monitoring networks using an ensemble clustering method in the three major metropolitan areas of Mexico. Atmos Pollut Res 11(8):1271–1280. https://doi.org/10.1016/j.apr.2020.05.005

Suris FNA, Abu Bakar MA, Ariff NM, Nadzir MSM, Ibrahim K (2022) Malaysia PM_{10} air quality time series clustering based on dynamic time warping. Atmosphere 13(4):503. https://doi.org/10.3390/atmos13040503

Van Egmond ND, Onderdelinden D (1981) Objective analysis of air pollution monitoring network data; spatial interpolation and network density. Atmos Environ (1967) 15(6):1035–1046. https://doi.org/10.1016/0004-6981(81)90104-9

Viqueira JRR, Villarroya S, Mera D, Taboada JA (2020) Smart environmental data infrastructures: bridging the gap between earth sciences and citizens. Appl Sci 10(3):856. https://doi.org/10.3390/app10030856

Virkus S, Garoufallou E (2019) Data science from a library and information science perspective. Data Technol Appl 53(4):422–441. https://doi.org/10.1108/DTA-05-2019-0076

Wang C, Zhao L, Sun W, Xue J, Xie Y (2018) Identifying redundant monitoring stations in an air quality monitoring network. Atmos Environ 190:256–268. https://doi.org/10.1016/j.atmosenv.2018.07.040

Wardana INK, Gardner JW, Fahmy SA (2021) Optimising deep learning at the edge for accurate hourly air quality prediction. Sensors 21(4):1064. https://doi.org/10.3390/s21041064

West M (1997) Time series decomposition. Biometrika 84(2):489–494. https://doi.org/10.1093/biomet/84.2.489
Wivou J, Udawatta L, Alshehhi A, Alzaabi E, Albeloshi A, Alfalasi S (2016) Air quality monitoring for sustainable systems via drone based technology. In: 2016 IEEE international conference on information and automation for sustainability (ICIAfS), 12/2016. IEEE, Galle, pp 1–5. https://doi.org/10.1109/ICIAFS.2016.7946542
Wu Q, Lin H (2019) A novel optimal-hybrid model for daily air quality index prediction considering air pollutant factors. Sci Total Environ 683:808–821. https://doi.org/10.1016/j.scitotenv.2019.05.288
Xu Y, Zhu Y, Shen Y, Yu J (2019) Fine-grained air quality inference with remote sensing data and ubiquitous urban data. ACM Trans Knowl Discov Data 13(5):46. https://doi.org/10.1145/3340847
Yang L, Wang J, Xie X, Kuang J (2013) Application of tucker decomposition in speech signal feature extraction. In: 2013 international conference on Asian language processing (IALP), 08/2013. IEEE, Urumqi, pp 155–158. https://doi.org/10.1109/IALP.2013.50
Yang L, Wang Y, Song H (2019) Research on data processing method of air quality monitoring system. In: 2019 4th international conference on mechanical, control and computer engineering (ICMCCE), 10/2019. IEEE, Hohhot, China, pp 452–4524. https://doi.org/10.1109/ICMCCE48743.2019.00108
Zabalza J, Ren J, Zheng J, Han J, Zhao H, Li S, Marshall S (2015) Novel two-dimensional singular spectrum analysis for effective feature extraction and data classification in hyperspectral imaging. IEEE Trans Geosci Remote Sens 53(8):4418–4433. https://doi.org/10.1109/TGRS.2015.2398468
Zammit-Mangion A (2023) Environmental data science: part 1. Environmetrics 34(1). https://doi.org/10.1002/env.2787
Zamora A, Dotta D, Chow Joe H, Tripathy RK, Paternina MRA (2019) Data-driven modal features extraction through the variational mode decomposition method. In: 2019 IEEE PES innovative smart grid technologies conference—Latin America (ISGT Latin America), 9/2019. IEEE, Gramado, pp 1–5. https://doi.org/10.1109/ISGT-LA.2019.8895351
Zhang L, Liu P, Zhao L, Wang G, Zhang W, Liu J (2021) Air quality predictions with a semi-supervised bidirectional LSTM neural network. Atmos Pollut Res 12(2):328–339. https://doi.org/10.1016/j.apr.2020.09.003
Zhu JY, Sun C, Li VOK (2017) An extended spatio-temporal granger causality model for air quality estimation with heterogeneous urban big data. IEEE Trans Big Data 3(3):307–319. https://doi.org/10.1109/TBDATA.2017.2651898

Chapter 2
Data Preprocessing in Air Quality Monitoring

Abstract This chapter explores the crucial steps of data preprocessing in air quality monitoring, focusing on missing value imputation and outlier detection. For missing value imputation, both univariate and multivariate methods are introduced, with examples of the former including linear interpolation, EM algorithms, and neural networks, and the latter including self-organizing maps (SOM) and multivariate nearest neighbor methods. Regarding outlier detection, the chapter discusses three approaches: unsupervised, filtering, and forecasting. Air quality data from the Jing-Jin-Ji region is utilized to analyze the temporal and spatial characteristics of pollutants, supporting the proposed data preprocessing methods. The analysis reveals that pollutant concentrations exhibit clear periodicity and spatial correlation.

2.1 Introduction

Air quality is an important concern for public health. To achieve effective air quality monitoring, the government has established many air pollution monitoring stations. These stations can collect air pollution data to support government decisions. However, due to equipment failures and noise, the collected data may contain missing values and outliers. These abnormal values will greatly damage the data information and reduce the application value of the model. To improve data quality and achieve effective monitoring of air quality, it is necessary to study data preprocessing methods including missing value completion and outlier detection.

In the actual environment, the data collected by many monitoring stations will be lost due to sensor failure, data transmission failure, etc. If these lost values are discarded directly, it will affect the distribution characteristics and correlation of the data. To rationally impute missing data, many researchers have proposed many methods of data missing data imputation. It can be divided into two categories, including univariate imputation methods and multivariate imputation methods, and so on:

H. Liu et al., *Data Science in Air Quality Monitoring*, Engineering Applications of Computational Methods 23, https://doi.org/10.1007/978-981-96-5777-3_2

- Common univariate completion methods can be divided into interpolation methods, EM methods, neural network methods, and so on. The interpolation methods can estimate missing data according to data trends. Kornelsen et al. utilized the linear interpolation method for missing data imputation (Kornelsen and Coulibaly 2014). The imputation performance is better than the benchmark methods. The Expectation-Maximization (EM) methods can analyze statistical characteristics and generate missing data imputation. Gold et al. proposed a novel structured-model EM method and proved the superiority of the EM method (Gold et al. 2003). The neural network methods can fit the correlation between the missing data and existing data. Gautam et al. designed a hybrid missing data imputation method based on a neural network (Gautam and Ravi 2015). In the proposed method, the clustering method and optimization method were applied to improve a naive neural network.
- The missing data imputation of a single variable will be affected by the length of the missing gap. For air pollution data, multivariate completion methods are very important because of the significant correlation between air quality variables. Depending on the correlation between multiple variables, it is expected to obtain better completion results. Currently commonly used multivariate data completion methods include Self-Organizing Map (SOM), multivariate nearest neighbor, etc. The SOM method can find the pattern of the multivariate data in the unsupervised manner and estimate missing data. Vatanen et al. used the SOM for missing data imputation, which outperforms the Principal Component Analysis (PCA) method. The multivariate nearest-neighbor method is a non-parametric method, which can calculate missing data according to the nearest rows. Tutz et al. utilized the nearest neighbor method and proved the effectiveness with larger predictors (Tutz and Ramzan 2015).

The air quality data may contain outliers due to systematic noise or failure. For air quality data, which has significant long-term dependency, outliers will greatly affect the analysis and prediction performance (Box et al. 2015). To improve data quality, many outlier detection methods are proposed, including unsupervised methods, filtering methods, forecasting-based methods, and so on:

- The unsupervised methods are essentially clustering methods. After clustering, the normal data and abnormal data are divided into different clusters. In this manner, the abnormal outliers can be separated from raw data. Liu et al. proposed the isolation forest method, which is similar to the random forest method in the aspect of algorithm principle (Liu et al. 2008). The isolation forest can detect outliers, given the fact that the abnormal data can be isolated with fewer binary trees. Çelik et al. applied the Density-Based Spatial Clustering of Applications with Noise (DBSCAN) method for outlier detection (Çelik et al. 2011). The DBSCAN is a mature clustering algorithm, which can cluster data according to density. With the DBSCAN, the abnormal data with low density can be detected.

- The filtering methods can analyze data and remove outliers according to time-frequency characteristics of the time series. Hampel identifier is one of the most widely-used outlier detection methods (Liu et al. 2004). The Hampel identifier is extended from the three-sigma rule and detects outliers with sliding windows. Kalman filter has been utilized for outlier detection (Ting et al. 2007). This outlier detection method can discover dynamic characteristics of the time series, and detect outliers.
- The forecasting-based methods can build a prediction model for the time series to describe the dynamic behavior. If the forecasting error is abnormal, the dynamic characteristics of the time series are different from others. Then, the outlier can be detected. The forecasting-based methods are widely-used in fault diagnosis. Zhang et al. utilized optimized neuron work for wind turbine diagnosis (Zhang et al. 2018). Kong et al. proposed a spatial-temporal model to detect outliers of the wind turbine condition data (Kong et al. 2020).

2.2 Data Acquisition

Jing-Jin-Ji area is one of the heavy industry centers in China, which contains 13 cities, including Beijing, Tianjin, Shijiazhuang, Zhangjiakou, Chengde, Qinhuangdao, Tangshan, Baoding, Langfang, Cangzhou, Hengshui, Xingtai and Handan. Because of the developed steel and coal industries, the air pollution of the Jing-Jin-Ji area is significantly severe. According to the Chinese air pollution report from January to June 2019, the Xingtai, Shijiazhuang, Handan, Baoding, and Tangshan in Jing-Jin-Ji area ranked 167, 166, 164, 161, and 157 among 168 important cities. Studying the air quality monitoring in Jing-Jin-Ji area is important for public health.

To ensure the universality of this study, 36,000 samples with 1-h time interval are used, which covers 2014-05-13 00:00 to 2018-06-21 00:00. Total seven air quality variables are analyzed in this study, including AQI, PM2.5, PM10, SO_2, NO_2, O_3 and CO. Each city has multiple monitoring stations. The pollutant concentrations of the cities are obtained by averaging the data of all stations. Taking Beijing as an example, the curve plot and violin plot of these air quality variables are shown in Fig. 2.1. It can be seen from Fig. 2.1a that all the pollutant time series show obvious periodic characteristics. In addition, there is a significant correlation between different pollutant data. For example, PM2.5 and PM10 have similar fluctuation characteristics. From Fig. 2.1b, it can be seen that the air quality data obey leptokurtic distribution with positive skewness. These characteristics indicate that the mean value of the air quality data is greater than the mode, and extreme values are more likely to occur than normal distributions.

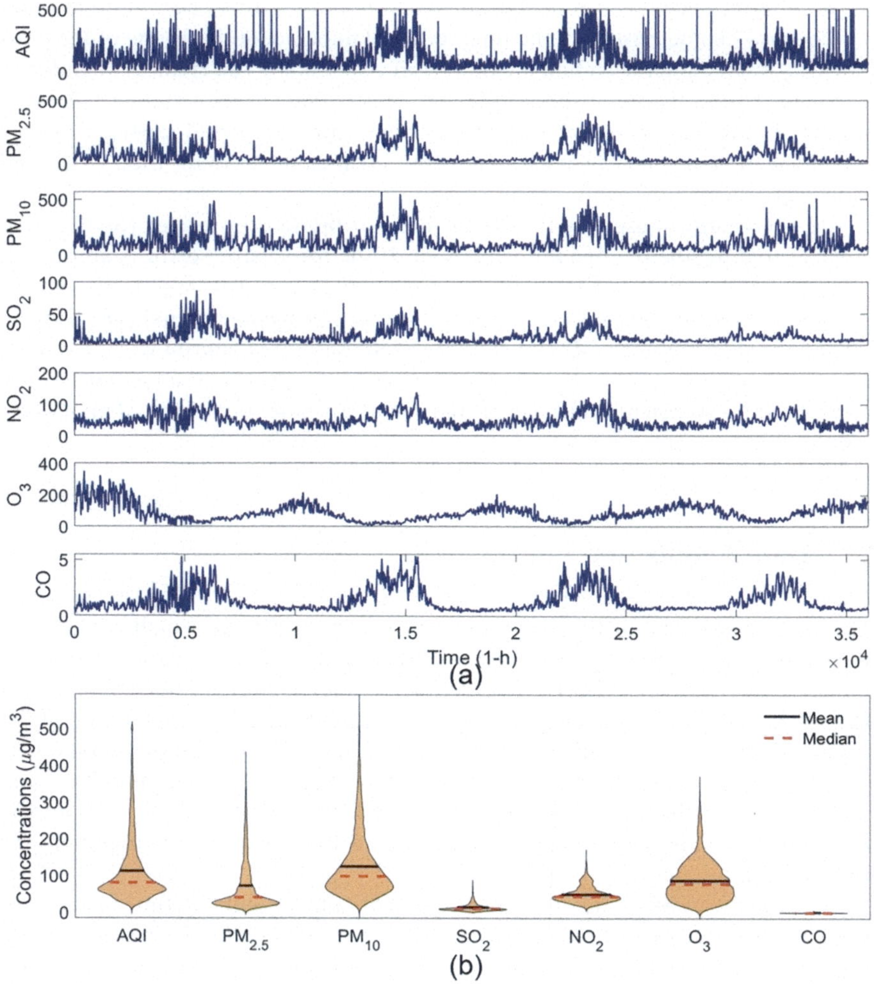

Fig. 2.1 The curve plots and violin plots of the air quality data in Beijing. (**a**) Curve plots. (**b**) Violin plots

2.3 Characteristic Analysis of Air Quality Data

2.3.1 Temporal Characteristics

2.3.1.1 Correlation Analysis Between Pollutants

The transformation between different pollutants is significant for air quality monitoring. To reveal the correlation between pollutants, the Spearman correlation index is applied in this section. The formula of the Spearman's correlation index p is presented as follows:

$$p = \frac{\sum_i (x_i - \bar{x})(y_i - \bar{y})}{\sqrt{\sum_i (x_i - \bar{x})^2 \sum_i (y_i - \bar{y})^2}} \tag{2.1}$$

where x and y are ranks of the analyzed variables. The advantage of the Spearman's correlation coefficient is that it does not require any prior. The Spearman's correlation coefficient can describe any form of correlation, while the Pearson correlation coefficient can only describe linear correlation.

The Spearman's correlation indexes between PM2.5 and other air quality variables in Beijing are presented in Fig. 2.2.

2.3.1.2 Periodicity Analysis

The frequency spectrum of the air quality data is analyzed to reveal the periodicity. In this study, the FFT is applied to analyze the frequency spectrum. Taking PM2.5 concentrations in Beijing as an example, the results are presented in Fig. 2.3. It can be seen from Fig. 2.3 that PM2.5 has a strong annual periodicity and semi-annual periodicity. This is because seasonal changes affect pollutant emissions and climatic

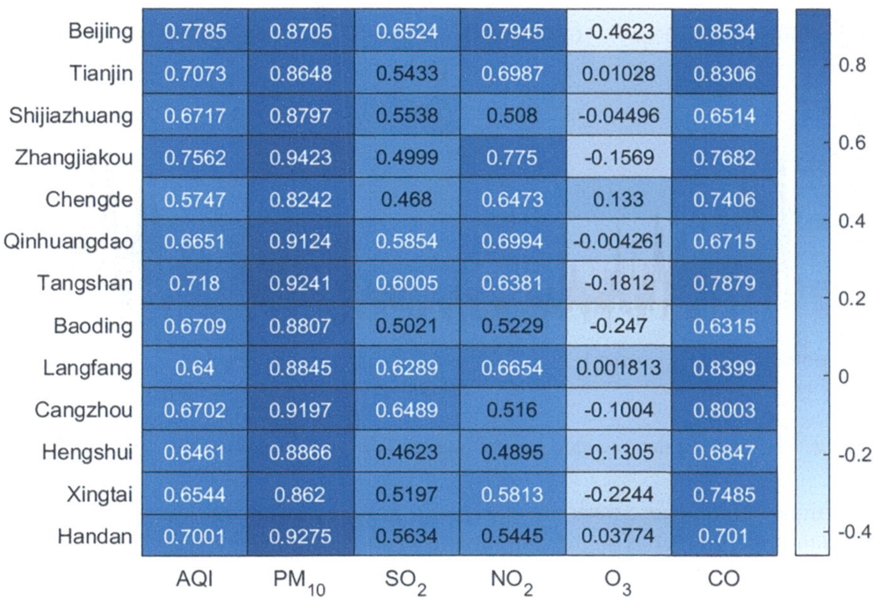

Fig. 2.2 The Spearman's correlation between PM2.5 and other variables in Beijing

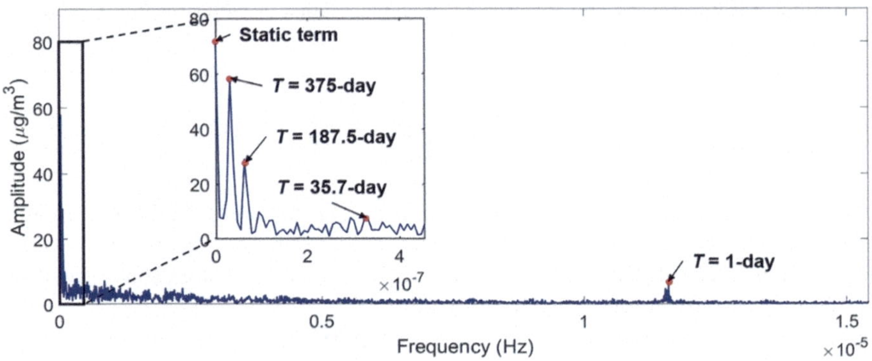

Fig. 2.3 The frequency spectrum of the PM2.5 concentrations series in Beijing

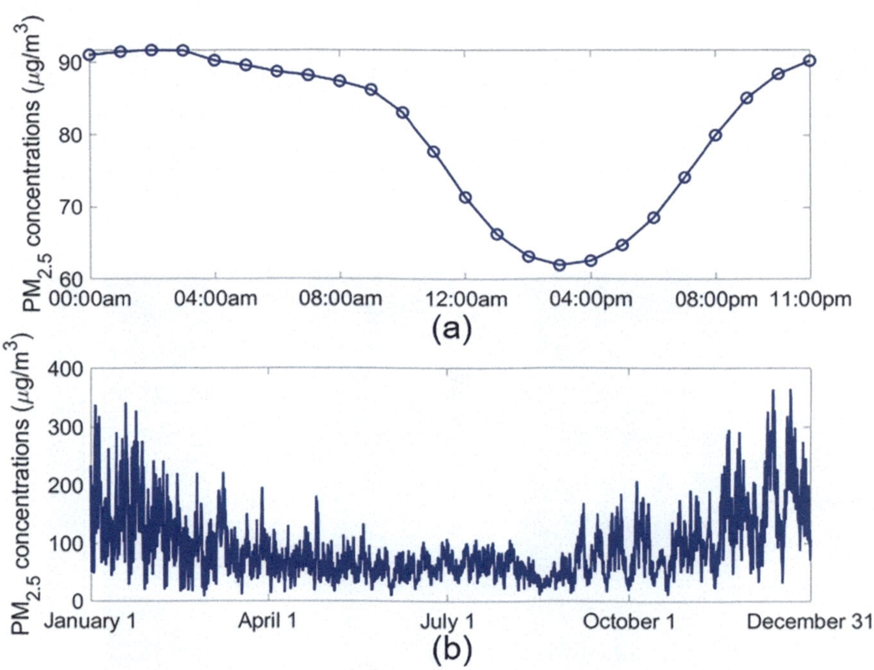

Fig. 2.4 The averaged trends within one day and one year. (**a**) One day.(**b**) One year

conditions, which cause periodic changes in pollutant concentrations. Besides, the pollutant time series has significant monthly and daily periodicity.

Taking daily and annual periodicities in Zhangjiakou as an example, the averaged trends within one day and one year are presented in Fig. 2.4. As can be seen from Fig. 2.4a, during the day, the PM2.5 concentration at 4 p.m. is the lowest and the concentration at night is the highest. This is caused by the inversion temperature.

The inversion temperature means that the lower air is cold, and the upper air is hot. In this case, the air can hardly flow up and down, and it is difficult for pollutants to diffuse, leading to increased concentration. At 4:00 p.m., the impact of the morning rush hour of traffic has dissipated, and the evening rush hour has not yet arrived. The near-surface atmospheric turbulence is strong, and the effect of temperature inversion is not obvious. At night, the ground temperature drops rapidly, and the upper air cools more slowly, forming the temperature inversion phenomenon and forming a peak of pollutants. It can be seen from Fig. 2.4b that pollutants are lower in summer and higher in other seasons. In the winter, because of the increase in coal combustion, the emission of pollutants is large. Moreover, the temperature inversion in winter is more obvious, which is not conducive to the diffusion of pollutants. There are more sand and dust storms in northern China in spring, and photochemical smog is prone to occur in autumn, so the PM2.5 concentration is higher. In summer, due to less temperature inversion and heavy rainfall, PM2.5 concentration is the lowest.

2.3.2 *Spatial Characteristics*

2.3.2.1 Spatial Correlation Analysis

The correlation of the air pollution between different cities is presented in Fig. 2.5. It can be seen from the figure that there is a strong correlation between different locations. This is due to the transmission of pollutants between cities.

2.3.2.2 Planar Maximally Filtered Graph Analysis

The Planar Maximally Filtered Graph (PMFG) is applied to analyze the local spatial correlation of the pollutant. Taking AQI as an example, the PMFG is shown in Fig. 2.6. After calculating the PMFG, the correlation between cities can be analyzed

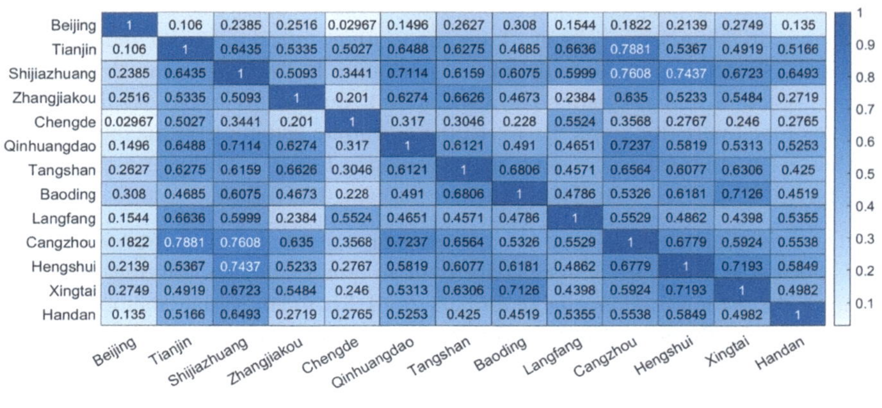

	Beijing	Tianjin	Shijiazhuang	Zhangjiakou	Chengde	Qinhuangdao	Tangshan	Baoding	Langfang	Cangzhou	Hengshui	Xingtai	Handan
Beijing	1	0.106	0.2385	0.2516	0.02967	0.1496	0.2627	0.308	0.1544	0.1822	0.2139	0.2749	0.135
Tianjin	0.106	1	0.6435	0.5335	0.5027	0.6488	0.6275	0.4685	0.6636	0.7881	0.5367	0.4919	0.5166
Shijiazhuang	0.2385	0.6435	1	0.5093	0.3441	0.7114	0.6159	0.6075	0.5999	0.7608	0.7437	0.6723	0.6493
Zhangjiakou	0.2516	0.5335	0.5093	1	0.201	0.6274	0.6626	0.4673	0.2384	0.635	0.5233	0.5484	0.2719
Chengde	0.02967	0.5027	0.3441	0.201	1	0.317	0.3046	0.228	0.5524	0.3568	0.2767	0.246	0.2765
Qinhuangdao	0.1496	0.6488	0.7114	0.6274	0.317	1	0.6121	0.491	0.4651	0.7237	0.5819	0.5313	0.5253
Tangshan	0.2627	0.6275	0.6159	0.6626	0.3046	0.6121	1	0.6806	0.4571	0.6564	0.6077	0.6306	0.425
Baoding	0.308	0.4685	0.6075	0.4673	0.228	0.491	0.6806	1	0.4786	0.5326	0.6181	0.7126	0.4519
Langfang	0.1544	0.6636	0.5999	0.2384	0.5524	0.4651	0.4571	0.4786	1	0.5529	0.4862	0.4398	0.5355
Cangzhou	0.1822	0.7881	0.7608	0.635	0.3568	0.7237	0.6564	0.5326	0.5529	1	0.6779	0.5924	0.5538
Hengshui	0.2139	0.5367	0.7437	0.5233	0.2767	0.5819	0.6077	0.6181	0.4862	0.6779	1	0.7193	0.5849
Xingtai	0.2749	0.4919	0.6723	0.5484	0.246	0.5313	0.6306	0.7126	0.4398	0.5924	0.7193	1	0.4982
Handan	0.135	0.5166	0.6493	0.2719	0.2765	0.5253	0.425	0.4519	0.5355	0.5538	0.5849	0.4982	1

Fig. 2.5 The correlation of the air pollution between different cities

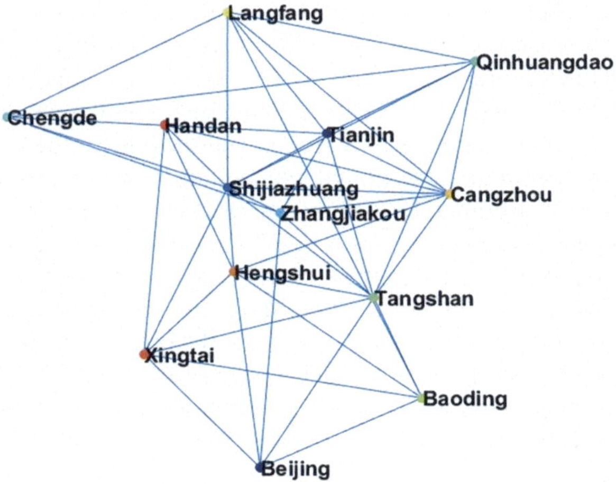

Fig. 2.6 The PMFG of the AQI

with graph theory. The maximum clique can represent the strongly correlated cities. For all pollution variables, the maximum cliques and their correlation indexes are presented in Table 2.1. It can be seen from Table 2.1 that Tianjin, Shijiazhuang, Qinhuangdao, and Cangzhou have the highest pollutant correlation.

2.4 Missing Data Imputation of Air Quality Data

There are three typical missing data patterns, including Missing Completely At Random (MCAR), Missing At Random (MAR), and Missing Not At Random (MNAR). The MCAR means the missed data are independent of the external variables. The MAR means the data are more likely to be missed in certain situations. The MNAR holds when the MAR and MCAR are not satisfied. As for the air quality data, the MAR is satisfied. The missing data of the air quality series contains two behaviors. In the missing behavior I, the monitoring station fails, and all air quality data are completely missed. In the missing behavior II, the monitoring station does not malfunction, and air quality data are randomly missed. To investigate the performance of the methods comprehensively, the univariate and multivariate methods are applied for these behaviors, which are presented in Fig. 2.7. In Fig. 2.7, the arrows represent the involved data. The details are explained as follows:

- The univariate method is applied to impute the missing data of behavior I, which is presented in Fig. 2.7a. In behavior I, all air quality data are missing. The multivariate method is not available for behavior I. So, the univariate method is used. As shown in the arrows in Fig. 2.7a, the data before and after the missing data are used for imputation.

Table 2.1 The maximum cliques and their correlation indexes

Air pollution variable	Maximum clique	Correlation index
AQI	[Tianjin, Shijiazhuang, Qinhuangdao, Cangzhou]	0.7127
PM2.5	[Tianjin, Shijiazhuang, Qinhuangdao, Cangzhou]	0.8085
PM10	[Tianjin, Shijiazhuang, Qinhuangdao, Cangzhou]	0.7770
SO_2	[Tianjin, Shijiazhuang, Tangshan, Cangzhou]	0.7941
NO_2	[Tianjin, Shijiazhuang, Qinhuangdao, Cangzhou]	0.7682
O_3	[Tianjin, Zhangjiakou, Tangshan, Cangzhou]	0.8577
CO	[Shijiazhuang, Tangshan, Hengshui, Xingtai]	0.7412

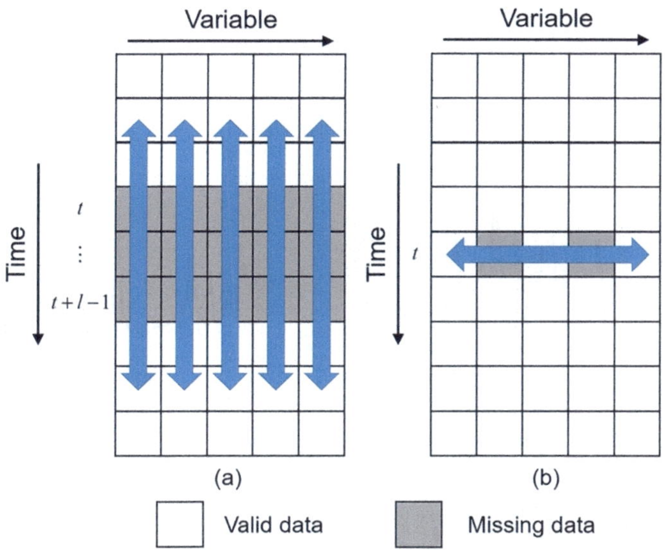

Fig. 2.7 The hybrid missing data imputation method. (**a**) Imputation method for behavior I. (**b**) Imputation method for behavior II

- The multivariate method is applied to impute the missing data of behavior II, which is presented in Fig. 2.7b. In behavior II, only part of air quality variable is lost. To study the relationship between different variables, the multivariate method is method. As shown in the arrow in Fig. 2.7b, the available data at the same time are used to impute the missing data. The advantage of the multivariate method is that no future information is used. So, it is feasible for online imputation.

2.4.1 Missing Data Imputation Performance Evaluation

To evaluate imputation performance fairly, three air quality series without missing data are extracted. The length of the series is set as 1500. These four data are captured from Shijiazhuang, Chengde, Cangzhou, and Handan respectively, as series #A1, A2, A3, and A4. To simulate the actual missing data behavior, the normal data in the air quality time series are replaced by the missing data randomly. The missing ratio is set as 5%, 10%, 15%, 20%, and 25% respectively to investigate the sensitivity of the imputation algorithm. The missing ratio between the first missing behavior and the second behavior is set as 9:1, which is similar to real behavior. Taking 25% missing ratio, the PM2.5 concentrations series are shown in Fig. 2.8. Taking series #1 as an example, the percentages of the missing gaps in the simulated series are presented in Table 2.2.

The imputation performance can be evaluated in the accuracy and stability. The accuracy can measure the difference between the estimated values and actual values. The stability can measure the fluctuating of the estimation error.

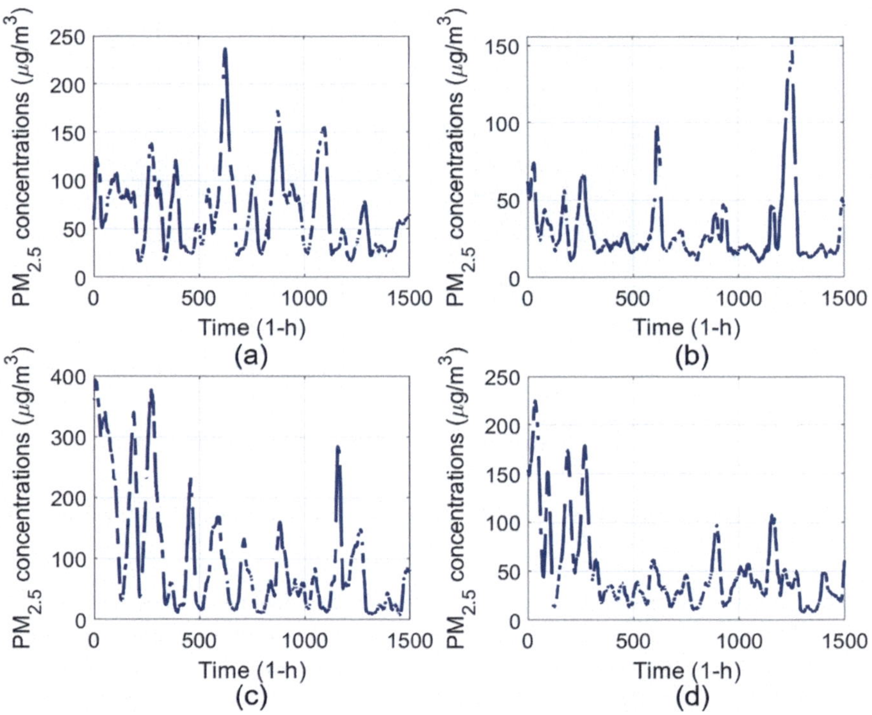

Fig. 2.8 The PM2.5 concentrations series with 25% missing ratio. (**a**) Series #A1. (**b**) Series #A2. (**c**) Series #A3. (**d**) Series #A4

Table 2.2 The percentages of the missing gaps in series #1

Missing ratio	$l \leq 1\,\text{h}$	$1\,\text{h} < l \leq 3\,\text{h}$	$l > 3\,\text{h}$
5%	95.59%	2.94%	1.47%
10%	84.92%	15.08%	0
15%	85.33%	14.67%	0
20%	82.02%	16.23%	1.75%
25%	75.38%	23.46%	1.15%

2.4.1.1 Accuracy

The commonly-used accuracy indexes contain Mean Averaging Error (MAE), Rooted Mean Square Error (RMSE), Pearson's correlation (P), and Kling–Gupta Efficiency (KGE). The MAE and RMSE are all based on the deviations. The MAE is suitable to describe the forecasting performance with uniform distribution, while the RMSE is suitable for normally distributed error (Chai and Draxler 2014). The smaller the MAE and RMSE, the better the accuracy. The P is a normalized index, which can measure the correlation between the actual values and estimated values. The larger the P, the better the accuracy. The KGE is a comprehensive evaluation index, which can consider both correlation and deviation. The KGE ranges between 1 and negative infinity. The larger the P, the better the accuracy.

The equations of the MAE, RMSE, P and KGE are presented as follows:

$$\text{MAE} = \left(\sum_{t=1}^{N} \left| Y_t - \hat{Y}_t \right| \right) / N \tag{2.2}$$

$$\text{RMSE} = \sqrt{\frac{1}{N} \sum_{t=1}^{N} \left(Y_t - \hat{Y}_t \right)^2} \tag{2.3}$$

$$\text{P} = \frac{\text{cov}\left(Y, \hat{Y} \right)}{\sigma_Y \sigma_{\hat{Y}}} \tag{2.4}$$

$$\text{KGE} = 1 - \sqrt{(P-1)^2 + \left(\frac{\sigma_{\hat{Y}}}{\sigma_Y} - 1 \right)^2 + \left(\frac{\mu_{\hat{Y}}}{\mu_Y} - 1 \right)^2} \tag{2.5}$$

where Y and $\hat{Y}$ are the actual and estimated values respectively; $\text{cov}\left(Y, \hat{Y} \right)$ is the covariance between the Y and $\hat{Y}$; σ_Y and $\sigma_{\hat{Y}}$ are the standard deviations of the Y and $\hat{Y}$; μ_Y and $\mu_{\hat{Y}}$ are the standard deviations of the Y and $\hat{Y}$.

2.4.1.2 Stability

The commonly used stability indexes contain Standard Deviation of Error (SDE), Dispersion Index of Error (DIE) and Interval Width of Error (IWE). The SDE and DIE can evaluate the fluctuating degree of the forecasting error. The SDE calculate the standard deviations, which the DIE calculates the ratio of the variance and mean. The IWE measures the width of confidence interval under α level. In this section, the α is set as 90% (Gneiting et al. 2007).

The equations of the SDE, DIE, and IWE are presented as follows:

$$\mathrm{SDE} = \sqrt{\frac{\sum_{t=1}^{N}\left(e_t - \overline{e}\right)^2}{N}} \tag{2.6}$$

$$\mathrm{DIE} = \frac{\sum_{t=1}^{N}\left(e_t - \overline{e}\right)^2}{\sum_{t=1}^{N}\left|e_t\right|} \tag{2.7}$$

$$\mathrm{IWE} = e_{95\%} - e_{5\%} \tag{2.8}$$

where e_t is the forecasting residual; $\overline{e}$ is averaging of the forecasting residuals; $e_{95\%}$ and $e_{5\%}$ are the 95% and 5% quantiles of the residual series.

2.4.2 *Univariate Missing Data Imputation Based on K-Nearest Neighbors*

2.4.2.1 Theoretical Basis

The K-Nearest Neighbors (KNN) method is a widely used algorithm to estimate univariate missing data. The biggest advantage of this method is its simplicity. Because the univariate imputation has a little available information, complex methods may overfit and generate incorrect imputation results. The theory of the KNN method is presented as follows:

$$\left[\hat{v}(t) \cdots \hat{v}(t+l)\right] = \left[v(t-k) \cdots v(t-1)\ v(t+l-1) \cdots v(t+l+k-1)\right] \begin{bmatrix} w_1(t-k) & \cdots & w_l(t-k) \\ \vdots & \cdots & \vdots \\ w_1(t+l+k-1) & \cdots & w_l(t+l+k-1) \end{bmatrix} \tag{2.9}$$

where l is the length of the missing gap, k is the number of the neighbours, $\hat{v}(t)$ is the imputed air quality data, $v(t)$ is the observed data, $w(t)$ is the weight for the observed data.

The weight matrix is the most important parameter for k-nearest neighbor method. In this chapter, the matrix is calculated by pseudo-inverse as follows:

$$\begin{bmatrix} w_1(t-k) & \cdots & w_l(t-k) \\ \vdots & \cdots & \vdots \\ w_1(t+l+k-1) & \cdots & w_l(t+l+k-1) \end{bmatrix} = \mathbf{V}_n^{\dagger}\mathbf{V}_i \tag{2.10}$$

where $\mathbf{V}_n^{\dagger} = \left[\mathbf{v}(t-k) \quad \cdots \quad \mathbf{v}(t-1) \quad \mathbf{v}(t+l-1) \quad \cdots \quad \mathbf{v}(t+l+k-1)\right]$ is a matrix composed of neighbor data, † is the pseudo-inverse, $\mathbf{V}_i = \left[\mathbf{v}(t) \quad \cdots \quad \mathbf{v}(t+l)\right]$ is a matrix composed of center data. Eq. (2.2) can infer the neighbor relationship in the series and calculate weights. Then, the missing values can be estimated according to the neighbor data as Eq. (2.1).

2.4.2.2 Modeling Step

The number of neighbors is an important parameter for the KNN method. The imputation performance is evaluated by P value. The P values of different numbers of neighbors are shown in Fig. 2.9. These values are obtained by five-fold cross validation. It can be observed that the missing data imputation performance is the best when the number of neighbors is equal to 2. The larger the number, the worse the performance. This phenomenon is because the two most recent data points contain the largest missing point information. Increasing the number of neighbors will increase model complexity and reduce generalization performance.

Figure 2.10 displays scatter plots comparing imputed and original data. The close proximity of the plotted points to the diagonal in Fig. 2.10 demonstrates the effectiveness of the proposed method in estimating missing data.

2.4.3 Multivariate Missing Data Imputation Based on Self-Organizing Map

2.4.3.1 Theoretical Basis

The SOM contains two layers, including an input layer and a competitive layer. The biggest feature of the SOM is that it is a non-supervised neural network. The SOM can map the competitive layer into the input layer with competitive learning. The

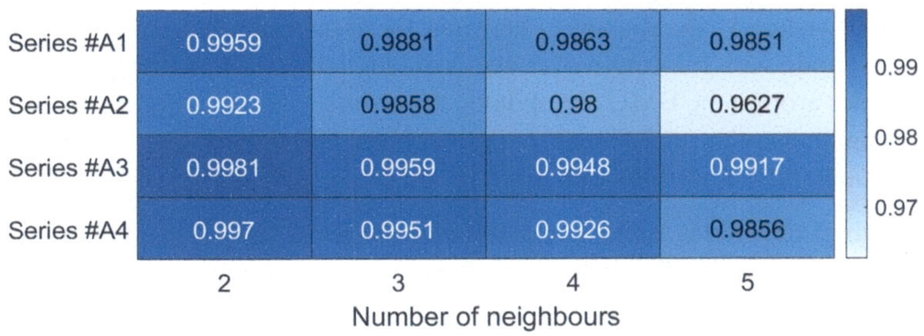

Fig. 2.9 The P values of different numbers of the neighbors

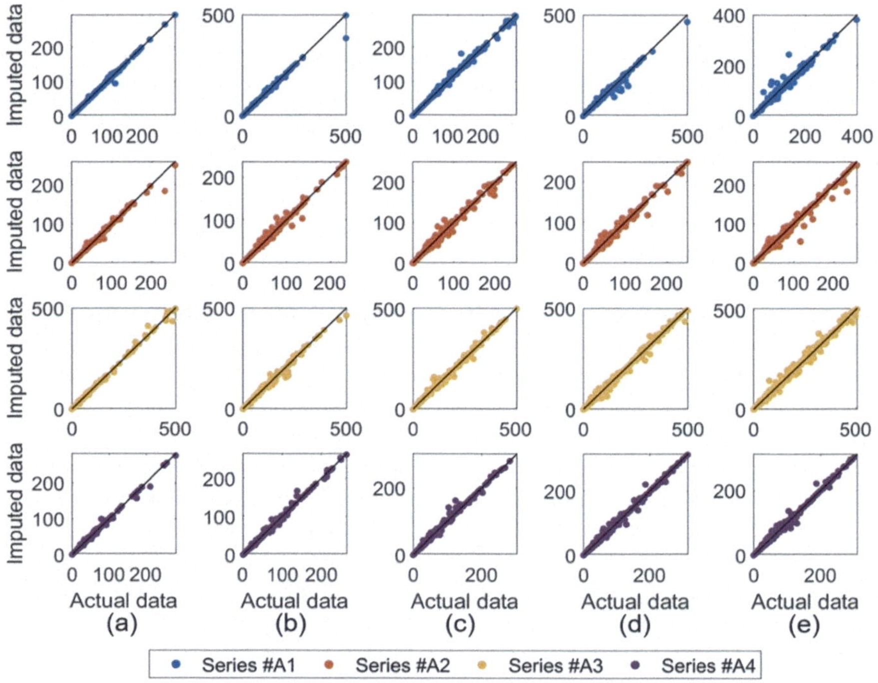

Fig. 2.10 The scatter plots of the univariate imputed data and original data. (**a**) 5% missing ratio. (**b**) 10% missing ratio. (**c**) 15% missing ratio. (**d**) 20% missing ratio. (**e**) 25% missing ratio

multivariate air quality vector is denoted as $[v_1, v_2, v_3, v_4, v_5, v_6, v_7]$, these variables represent AQI, PM2.5, PM10, SO_2, NO_2, O_3 and CO respectively. By feeding the multivariate air quality data into the SOM, the nodes in the competitive layer are trained to represent the pattern of the air quality data. The training process of the SOM can be explained as follows:

1. Initialize completive layer: The 2-D completive layers are initialized by many nodes. The typology of these nodes can be divided into hexagonal grid and rectangular grid, which are presented in Fig. 2.11.
2. For input vector V, the distances between the input and all nodes as calculated. The best node with minimum distance is denoted as Best Matching Unit (BMU).
3. The nodes around the BMU are updated according to the neighborhood function. The neighbor nodes are learned to approach the BMU.
4. Continue iteration until the condition is satisfied.

After training, the nodes represent pattern of the air quality data. Given a vector with missing element V_m, the distances between the vector and nodes are calculated. The missing element is ignored when computing the Euclidean distance. With the

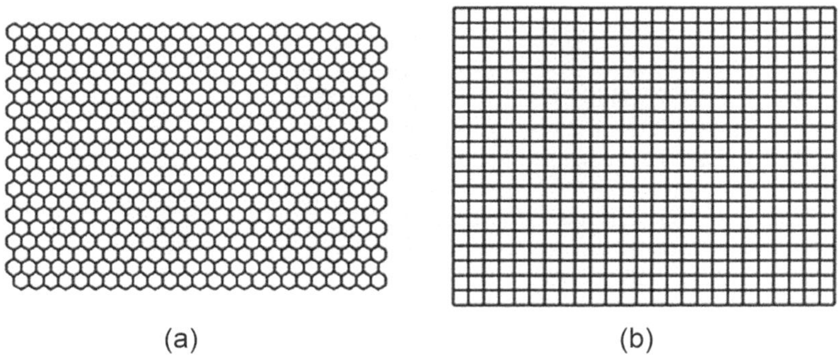

Fig. 2.11 The hexagonal grid and rectangular grid of the SOM. (a) Hexagonal grid. (b) Rectangular grid

BMU, the nearest complete vector can be found. Then, the missing data can be estimated with the nearest vector.

2.4.3.2 Modeling Step

To ensure the fairness of the analysis, the map size of the SOM model is tuned. The graph size can control how well the SOM model fits the data. If the size is too small, it will lead to incomplete fitting of the data information. But if the size is too large, the model will fall into overfitting. The P values of different map sizes are shown in Fig. 2.12. These values are obtained by five-fold cross validation. It can be observed that the missing data imputation performance is the best when the the maps sizes are equal to 95, 85, 75 and 95 respectively.

The complete air quality data are inputted into the SOM model to calculate nodes. Taking series #1 as an example, the initialized input and output spaces are shown in Figs. 2.13 and 2.14. It can be observed that the initial input space of the SOM is far from the output space, while the trained input space can approach the output space.

With the trained input space, the incomplete vector is inputted into SOM, and the corresponding BMU can be found. The actual data and the estimated data are presented in Fig. 2.15. From Fig. 2.15, it can be observed that the estimated data is close to the real values, indicating good performance of the SOM.

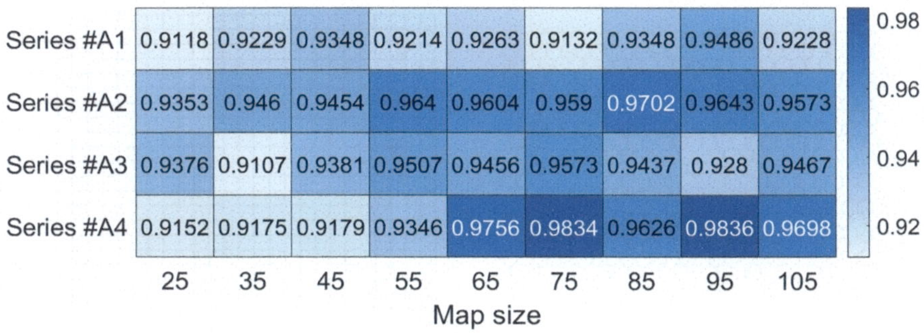

Fig. 2.12 The P values of different map sizes

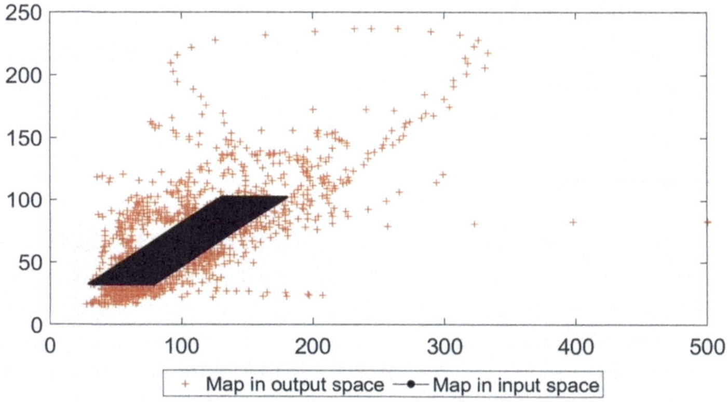

Fig. 2.13 The initialized input and outpue spaces of the SOM

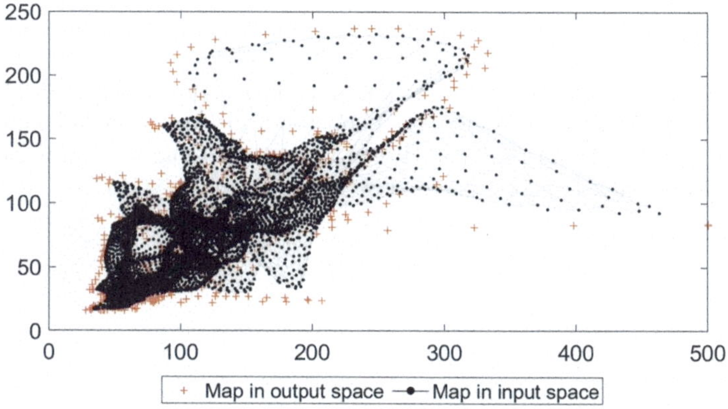

Fig. 2.14 The input and outpue spaces of the SOM after training

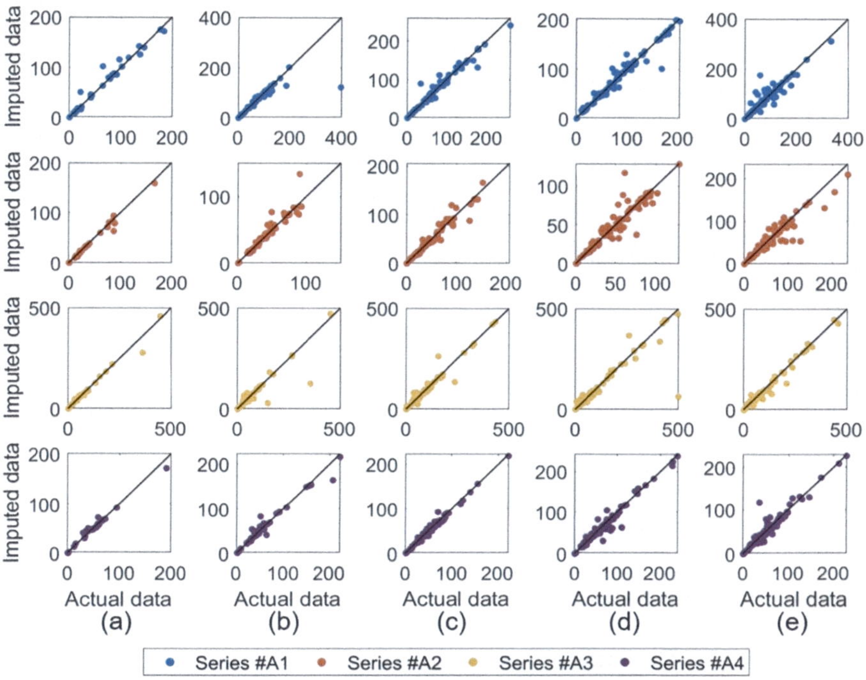

Fig. 2.15 The scatter plots of the multivariate imputed data and original data. (**a**) 5% missing ratio. (**b**) 10% missing ratio. (**c**) 15% missing ratio. (**d**) 20% missing ratio. (**e**) 25% missing ratio

2.5 Outlier Detection of Air Quality Data

In this chapter, three different outlier detection methods are described, including an unsupervised method, a filtering method, and a forecasting method. The mechanisms of these studied methods are shown in Fig. 2.16. As shown in Fig. 2.16a, the unsupervised outlier detection method can describe the correlation-ship between different variables. The variables with abnormal behavior can be found as outliers. As shown in Fig. 2.16b, the filtering method can fit the abnormal temporal pattern. As shown in Fig. 2.16c, the forecasting method is also a temporal outlier detection method. Different with the filtering method, the forecasting method only uses previous data for each point. These characteristics make the forecasting-based outlier detection method can be applied online.

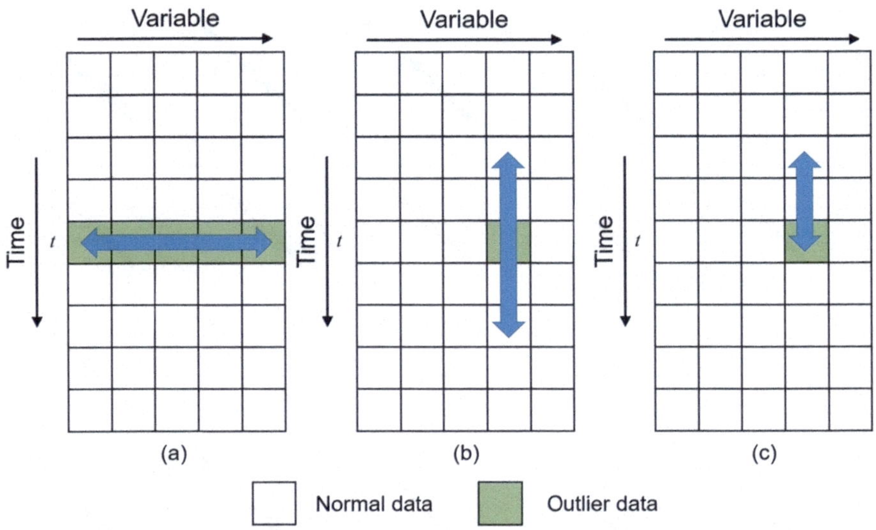

Fig. 2.16 The mechanisms of the outlier detection methods. (**a**) Clustering method. (**b**) Flitering method. (**c**) Forecasting method

2.5.1 Outlier Detection Performance Evaluation

Because the studied air quality data are not artificial, the real outliers cannot be obtained. Therefore, the real outlier detection accuracy cannot be calculated. In this chapter, the outlier detection performance is evaluated with the help of the forecasting algorithm. A well-preformed anomaly detection algorithm can eliminate abnormal information in the data and improve the prediction accuracy of the model. To ensure the fairness, 4 sets of air quality data from different sites and times are selected for the case study. Series #B1 is collected from 2015-01-01 00:00 to 2015-12-26 23:00 in Beijing; series #B2 is collected from 2015-06-01 00:00 to 2016-05-25 23:00 in Tianjin; series #B3 is collected from 2016-01-01 00:00 to 2017-12-25 23:00 in Shijiazhuang; series #B4 is collected from 2016-06-01 00:00 to 2017-05-26 23:00 in Zhangjiakou. Each dataset covers 8640 h. The missing data in these series are imputed with the KNN method. Taking PM2.5 as an example, the series are shown in Fig. 2.17.

After the outliers are detected, they are replaced via univariate missing data imputation algorithm in Sect. 2.4. Then, the processed series is used to train a Multi-Layer Perceptron (MLP) model. The trained model is used to predict another series with outliers. If the outlier detection algorithm can generate good outlier detection results, the forecasting model can fit the autocorrelation function within the series and obtain accurate forecasting results. Otherwise, the forecasting results become worse. In this section, the 1st ~ 7200th of the studied data are used for outlier detection and training models, and the trained MLP model is verified on the 7201st ~ 8640th data. The forecasting performance is evaluated with the indexes in Sect. 2.4.1.

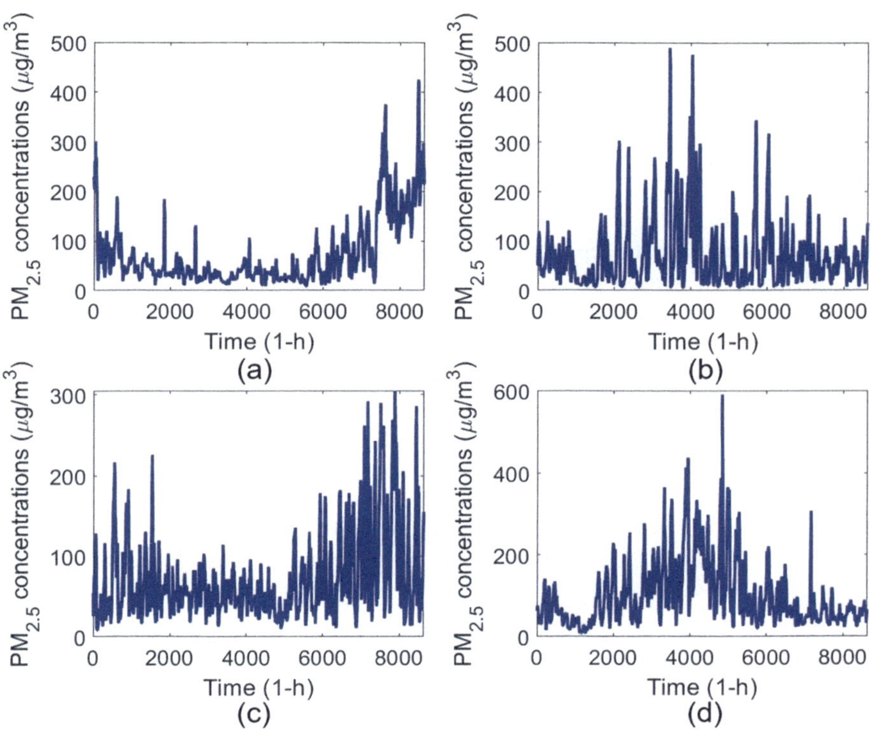

Fig. 2.17 The PM2.5 concentrations series with outliers. (**a**) Series #B1. (**b**) Series #B2. (**c**) Series #B3. (**d**) Series #B4

2.5.2 Outlier Detection Based on Unsupervised Isolation Forest

The air quality series contains many variables. The correlation between these variables can reflect the patterns of the air quality. The unsupervised method can learn the difference between the patterns and obtain the abnormal pattern.

With the unsupervised outlier detection methods, the isolation forest has the advantage of high computational efficiency. As a popular algorithm, the isolation forest has been widely applied for time series outlier detection. Lin et al. applied the isolation forest to improve the forecasting accuracy and proved the superiority of the isolation forest (Lin et al. 2020). Qin et al. combined the isolation forest algorithm and clustering method and proposed an improved outlier detection method for hydrology time series (Qin and Lou 2019). In this chapter, the isolation forest is used for multivariate outlier detection.

2.5.2.1 Theoretical Basis

The isolation forest is based on the theory of the random forest. The isolation forest algorithm separates the sample space by the isolation trees. Then, the outliers can be detected according to the anomaly score. The details are presented as follows (Qin and Lou 2019):

1. Construct isolation tree. According to the training samples X', a tree is constructed randomly to separate the attribution q of the samples X'. The samples with larger attributions are assigned into the right nodes, while the samples with smaller attributions are assigned into the left nodes. The tree are subsequently constructed iteratively on the right and left nodes until the data cannot be separated or the tree is deep enough.
2. Construct isolation forest. Setting the ensemble number is ψ, the ψ trees are constructed iteratively and combined to the forest. To reduce the similarity of the trees, the training samples X' are extracted from the whole sample set X for each tree.
3. Calculate anomaly score. For each data sample, the path length $h(x)$ is calculated as the length from the root to the node. The score can be calculated as follows:

$$S(x,n) = 2^{-\frac{h(x)}{c(n)}} \tag{2.11}$$

where $c(n)$ is the normalization parameter, which can be calculated as follows:

$$c(n) = 2H(n-1) - \left(\frac{2n-2}{n}\right) \tag{2.12}$$

where $H(n-1)$ is the harmonic number, which is equal to $\ln n + 0.5772156649$.

4. Select outliers. According to the anomaly score, the outliers can be detected. The sample with large anomaly scores are more likely to be an outlier, and verse visa.

2.5.2.2 Modeling Step

For each time, seven variables can be observed, including AQI, PM2.5, PM10, SO_2, NO_2, O_3, and CO. With the isolation forest, these data are used as features to represent the characteristics of the air quality. The multivariate air quality vector is denoted as $[v_1, v_2, v_3, v_4, v_5, v_6, v_7]$, these variables represent AQI, PM2.5, PM10, SO_2, NO_2, O_3, and CO respectively. These data are fed into the isolation for calculating the anomaly scores. The scatter plots of these data are presented in Fig. 2.18. The t-SNE method is applied to visualize the high-dimension air quality data. The color of the scatters represents the anomaly score. It can be seen from Fig. 2.18 that the points from the edge of the scatter group have higher scores, that is, the more likely they are outliers.

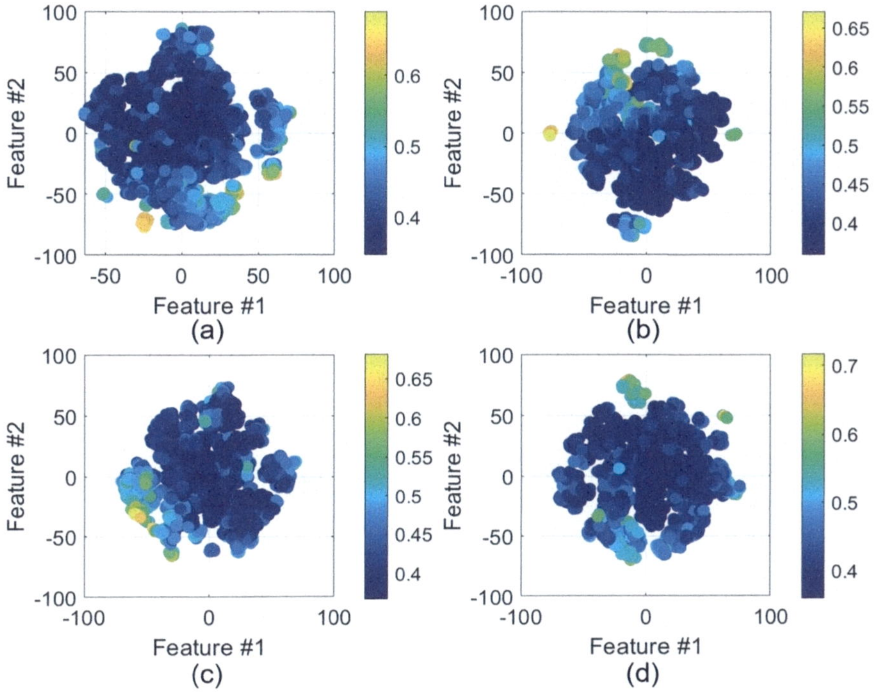

Fig. 2.18 The scatter plots of the air quality data. (**a**) Series #B1. (**b**) Series #B2. (**c**) Series #B3. (**d**) Series #B4

The histograms of the scores are presented in Fig. 2.19. As can be seen from Fig. 2.19, the scores have significant skewed distributions. Only a few points have large scores.

In the isolation forest algorithm, contamination is an important parameter that can decide the amount of outliers. To ensure fairness, the parameter is selected by cross-validation. Different from the missing data imputation model, the validation of the forecasting model should avoid dependency on samples. So, the cross-validation is achieved by five-fold blocked cross-validation. The difference between classical cross-validation and blocked cross-validation is shown in Fig. 2.20.

To avoid data leakage, the validation is carried out in the training data (1st ~ 7200th data). The default parameter of the contamination is 0.1. The research range of the contamination is set as [0.02, 0.04, …, 0.18, 0.20]. The P of the prediction model is used for performance evaluation. The P values with different contaminations are presented in Fig. 2.21, where the P values are obtained by averaging all air quality variables. From Fig. 2.21, it can be observed that the contamination value is important for the outlier detection. For series #B1, B2, B3, and B4, the optimal contamination values are 0.04, 0.02, 0.02, and 0.08 respectively.

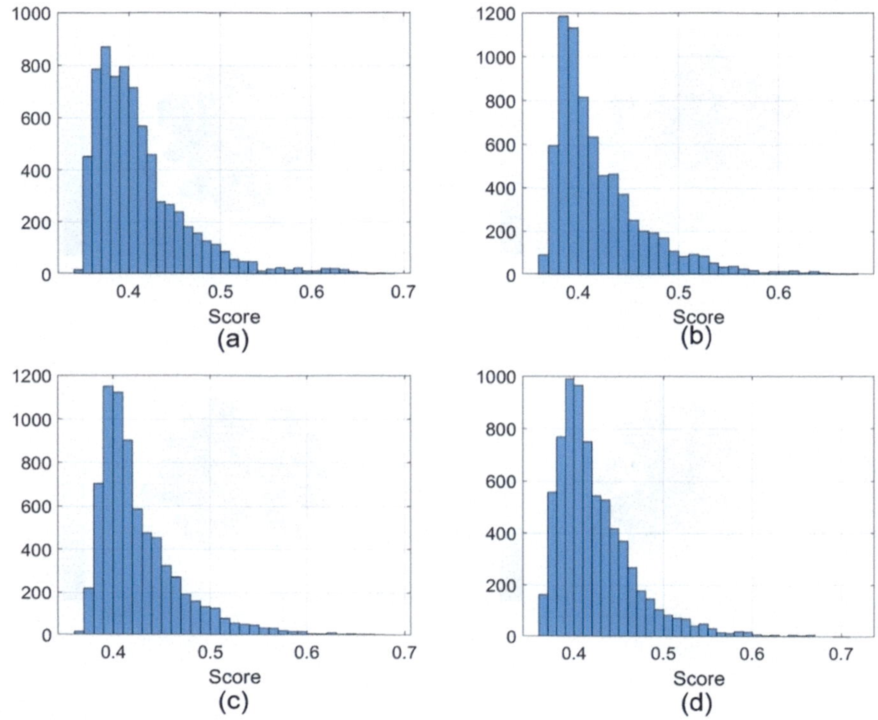

Fig. 2.19 The anomaly scores of the air quality data. (**a**) Series #B1. (**b**) Series #B2. (**c**) Series #B3. (**d**) Series #B4

With the obtained contamination value, the outliers are detected. These detected outliers are regarded as missing data and are imputed with the KNN method. Taking the PM2.5 series in series #B1 as an example, the detected outliers are shown in Fig. 2.22.

With the filtered air quality series in 1st ~ 5760th h, an MLP model can be trained to predict data in 5761st ~ 7200th h. Figure 2.23 displays scatter plots comparing the actual values and predicted values. From Fig. 2.23, it can be observed that the scatters are close to the diagonal. This phenomenon indicates the prediction model after the isolation forest can effectively estimate the missing data.

2.5.3 Outlier Detection Based on Hampel Filter

The filtering algorithm can suppress specific frequency components in the sequence. Because the outliers mainly contain high-frequency components, low-pass filtering algorithms are needed for processing. In this chapter, the widely used Hampel filter

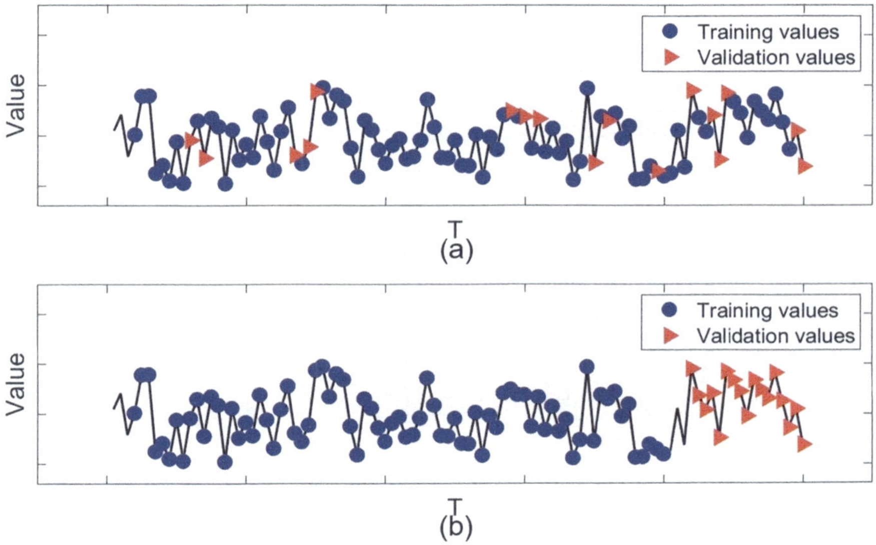

Fig. 2.20 The difference between classical cross-validation and blocked cross-validation. (**a**) Classical cross-validation. (**b**) Blocked cross-validation

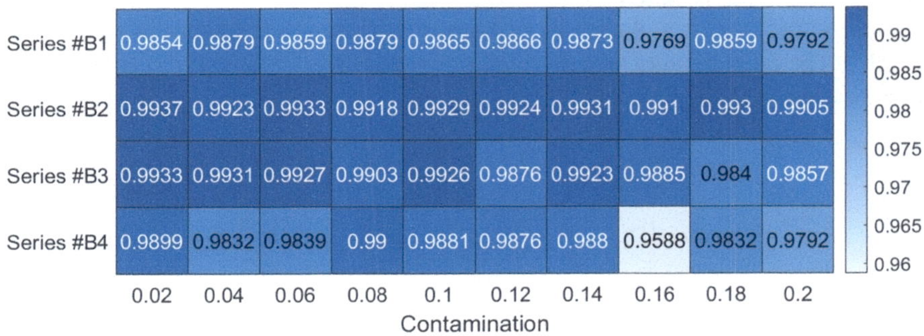

Fig. 2.21 The P values with different contaminations

is applied for outlier detection. Ghaleb et al. applied the Hampel filter to eliminate outliers in the electrocardiogram signal (Ghaleb et al. 2018). Sharadge et al. applied the Hampel filter to improve forecasting performance (Sharadga et al. 2020). Liu et al. used the Hampel filter to preprocess the PM2.5 time series, and built an ensemble model for prediction (Liu et al. 2019). The results indicate the effectiveness of the Hampel filter.

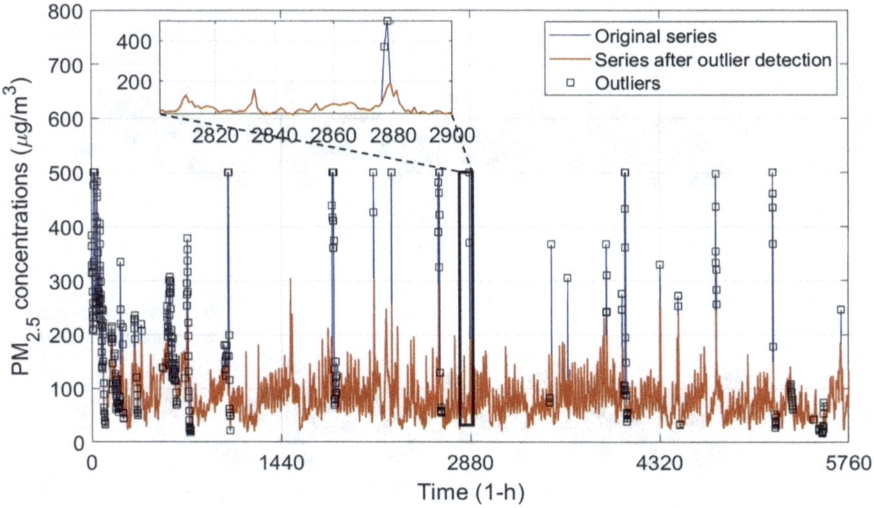

Fig. 2.22 Outlier detection results of the isolation forest

2.5.3.1 Theoretical Basis

The Hampel filter is based on the three-sigma criterion. Assuming the series is denoted as $[x_1, x_2, x_3, \ldots]$, the series is divided by sliding windows with single-side length l. The local median m_i and Median Absolute Deviation (MAD) σ_i for detection are estimated as follows:

$$m_i = median\left(x_{i-l}, x_{i-l+1}, \ldots, x_{i+l-1}, x_{i+l}\right) \tag{2.13}$$

$$\sigma_i = \kappa median\left(\left|x_{i-l} - m_i\right|, \left|x_{i-l+1} - m_i\right|, \ldots, \left|x_{i+l-1} - m_i\right|, \left|x_{i+l} - m_i\right|\right) \tag{2.14}$$

where $\kappa \approx 1.4826$.

The Hampel identifier can detect whether the sample x_i is the outlier. The detection criterion is presented as follows:

$$Z_i = \frac{\left|x_i - m_i\right|}{\kappa \sigma_i} \tag{2.15}$$

If the Z_i is larger than a threshold TR, then the sample is detected as an outlier.

2.5.3.2 Modeling Step

In the Hampel filter, the sliding window length $2l + 1$ is important. The length determines the trade-off between the local characteristics and global characteristics. In this chapter, the sliding window length is selected by cross-validation. The research range of the sliding window length is set as [3, 5, …, 23, 25]. The P of the prediction model is used for performance evaluation. The P values with different sliding window lengths are presented in Fig. 2.24, where the P values are obtained by averaging all air quality variables. From Fig. 2.24, it can be observed that the sliding window length is important for outlier detection. For series #B1, B2, B3, and B4, the optimal window lengths are 5, 19, 21, and 5 respectively.

With the obtained sliding window length, the outliers are detected according to the local median and MAD. Taking the PM2.5 series in series #B1 as an example, the detected outliers are shown in Fig. 2.25.

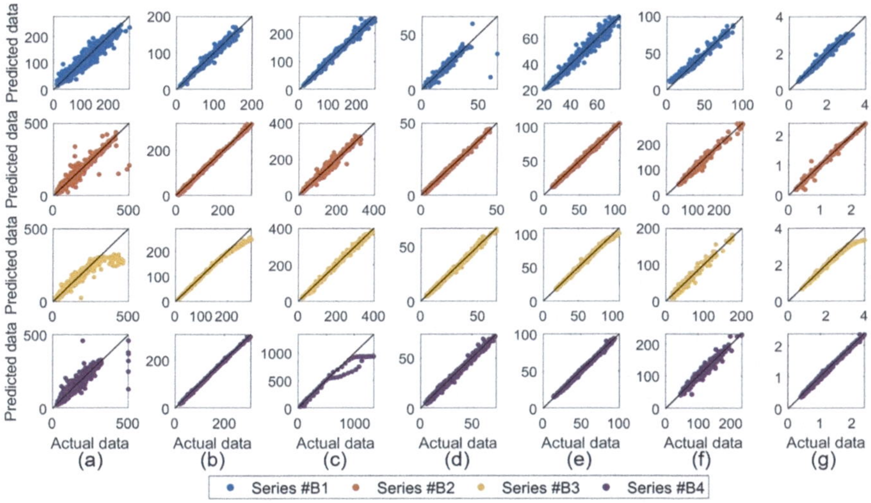

Fig. 2.23 The scatter plots of the predicted data and original data after isolation forest. (**a**) AQI. (**b**) PM2.5. (**c**) PM10. (**d**) SO_2. (**e**) NO_2. (**f**) O_3. (**g**) CO

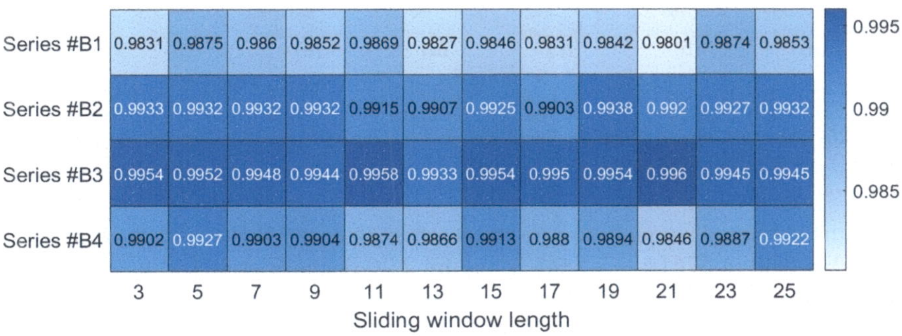

Fig. 2.24 The P values with different sliding window lengths

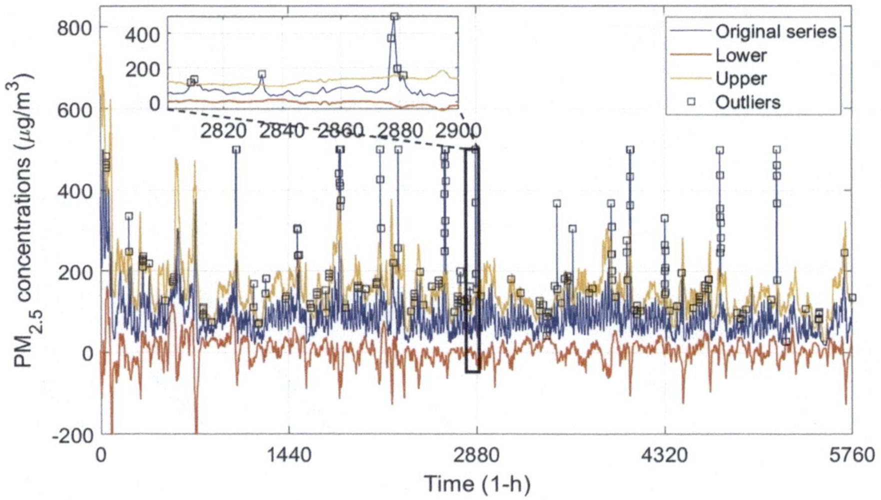

Fig. 2.25 Outlier detection results of the Hample filter

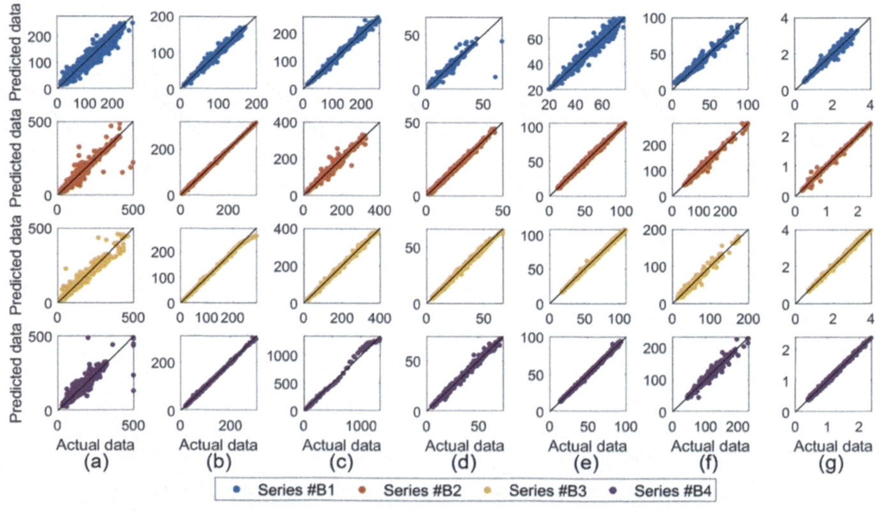

Fig. 2.26 The scatter plots of the predicted data and original data after Hampel filter. (**a**) AQI. (**b**) PM2.5. (**c**) PM10. (**d**) SO_2. (**e**) NO_2. (**f**) O_3. (**g**) CO

With the filtered air quality series in 1st ~ 5760th h, an MLP model can be trained to predict data in 5761st ~ 7200th h. Figure 2.26 displays scatter plots comparing the actual values and predicted values. From Fig. 2.26, it can be observed that the scatters are close to the diagonal. This phenomenon indicates the prediction model after the Hamel filter can effectively estimate the missing data.

2.5.4 Outlier Detection Based on Deep Learning Forecasting

For each air quality variable, the temporal autocorrelation function can reflect the pattern of the time series. Because of the non-linear and heteroscedasticity characteristics of the air quality series, the deep learning-based quantile regression method is applied for outlier detection.

2.5.4.1 Theoretical Basis

The forecasting accuracy algorithm can describe the dynamic behavior of the air quality series. If the dynamic behavior is not changed, the forecasting error remains stable. If the error is abnormal, the outlier occurs. In this chapter, the forecasting is achieved by decomposition-forecasting structure. The decomposition can separate the raw air quality series into several subseries. This subseries is more stationary than the original series. For each series, an individual model is applied for forecasting.

Assuming the original series is denoted as $\{x_t\}$, the decomposed subseries are $\left\{x_t^1\right\}$, $\left\{x_t^2\right\}$, …, $\left\{x_t^N\right\}$, where N is the number of the subseries. The decomposition is carried out with the Wavelet Packet Decomposition (WPD) algorithm. The algorithm can divide the raw series via a complete binary tree. For each subseries, the Long Short Term Memory (LSTM) algorithm is built for prediction. The LSTM model is proposed to solve the gradient explosion and gradient vanishing problem of the typical recurrent neural network. Figure 2.27 illustrates the detailed contents of an LSTM cell (LeCun et al. 2015).

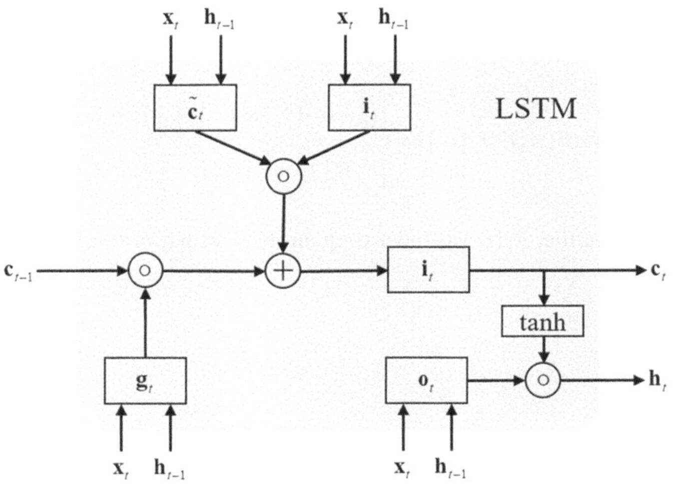

Fig. 2.27 The detail of an LSTM cell

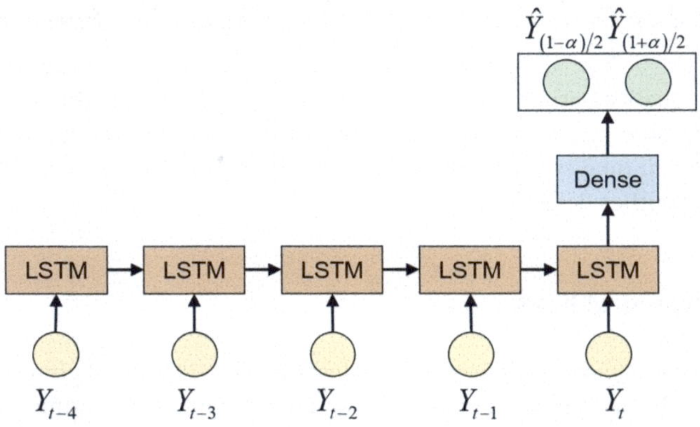

Fig. 2.28 The structure of the quantile regression LSTM network

This study employs a three-layer perceptron as the quantile regression network. The network processes historical data $[Y_t, Y_{t-1}, \ldots, Y_{t-4}]$ to output quantile regression results. Assuming the confidence level is α. Then, the $(1 - \alpha)/2$ and $(1 + \alpha)/2$ quantile regression results are required. The architecture of the quantile regression LSTM network is detailed in Fig. 2.28.

The neural network's training loss function integrates the Median Absolute Deviation losses for both quantiles. These losses are combined with equal weights of one. The quantile regression network's loss formula is presented below:

$$Loss = \sum_{\alpha \in \{0.5\%, 99.5\%\}} \frac{1}{N} \sum_{i=1}^{N} pinball\left(\alpha, Y, \hat{Y}_{\alpha}\right) \tag{2.16}$$

where Y is the actual data, $\hat{Y}_{\alpha}$ is the forecasting results of α-th quantile, pinball loss $pinball\left(\alpha, Y, \hat{Y}_{\alpha}\right)$ is defined as follows:

$$pinball\left(\alpha, Y_t, \hat{Y}_t(\alpha)\right) = \begin{cases} (1-\alpha)\left(\hat{Y}_t(\alpha) - Y_t\right) & Y_t < \hat{Y}_t(\alpha) \\ \alpha\left(Y_t - \hat{Y}_t(\alpha)\right) & Y_t \geq \hat{Y}_t(\alpha) \end{cases} \tag{2.17}$$

where α is the quantile, $\hat{Y}_t(\alpha)$ is the α-th quantile forecasting results, Y_t is the actual wind speed data.

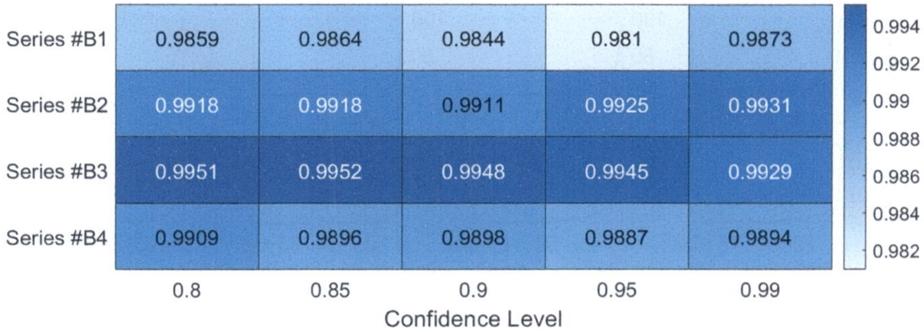

Fig. 2.29 The P values with different confidence levels

2.5.4.2 Modeling Step

In the deep learning forecasting method, the confidence level α should be fine-tuned. If the confidence level is too high, the sensitivity to outliers will be too low. If the confidence is too low, the normal value will be falsely detected as an outlier. In this chapter, the confidence level is selected by cross-validation. The research range of the sliding window length is set as [0.80, 0.85, 0.90, 0.95, 0.99]. The P values with different confidence levels are presented as in Fig. 2.29, where the P values are obtained by averaging on all air quality variables. From Fig. 2.29, it can be observed that the confidence level is important for outlier detection. For series #B1, B2, B3 and B4, the optimal confidence levels are 0.99, 0.99, 0.85 and 0.80 respectively.

With the obtained confidence levels, the LSTM can be trained via the Adam optimizer. The quantile regression loss during training is depicted in Fig. 2.30. As shown in Fig. 2.30, the Adam optimizer effectively minimizes errors. The error curve drops sharply at the beginning and then tends to be stable, indicating that the optimization has good convergence performance. In all the series and air quality variables, the Adam optimizer achieves robust optimization performance, highlighting the algorithm's reliability.

The trained LSTM model can generate upper and lower bounds. The outliers are detected as the samples beyond bounds. Taking the PM2.5 series in series #B1 as an example, the detected outliers and the imputed values are shown in Fig. 2.31.

With the filtered air quality series in 1st ~ 5760th h, an MLP model can be trained to predict data in 5761st ~ 7200th h. The scatter plots of the actual values and predicted values are shown in Fig. 2.32. From Fig. 2.32, it can be observed that the scatters are close to the diagonal. This phenomenon indicates the prediction model after the LSTM can effectively estimate the missing data.

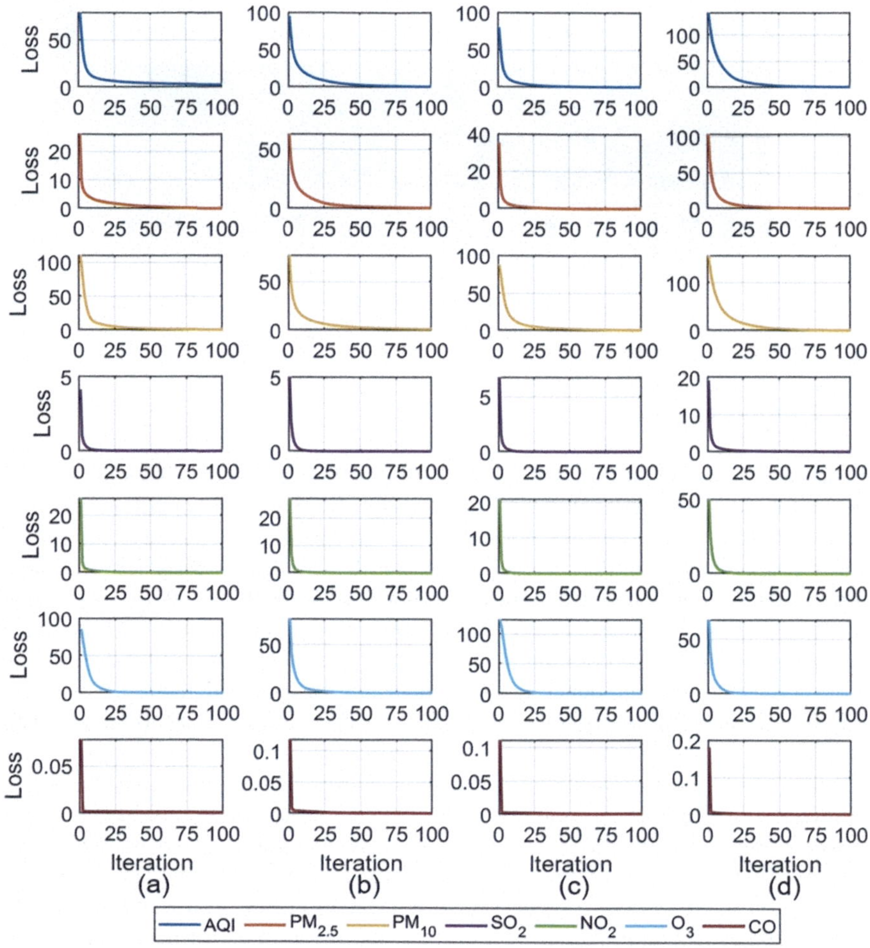

Fig. 2.30 The quantile regression loss during the LSTM's training. (**a**) Series #B1. (**b**) Series #B2. (**c**) Series #B3. (**d**) Series #B4

2.6 Preprocessing Performance Comparison

2.6.1 Performance Comparison of Missing Data Imputation

In this chapter, the univariate and multivariate missing data imputation methods are applied for different missing patterns. Taking 25% missing ratio as an example, the missing data imputation performance of the univariate and multivariate missing data imputation methods is presented in Table 2.3. The relation between the imputation performance and the missing ratio is shown in Fig. 2.33.

From Table 2.3 and Fig. 2.33, it can be concluded as follows:

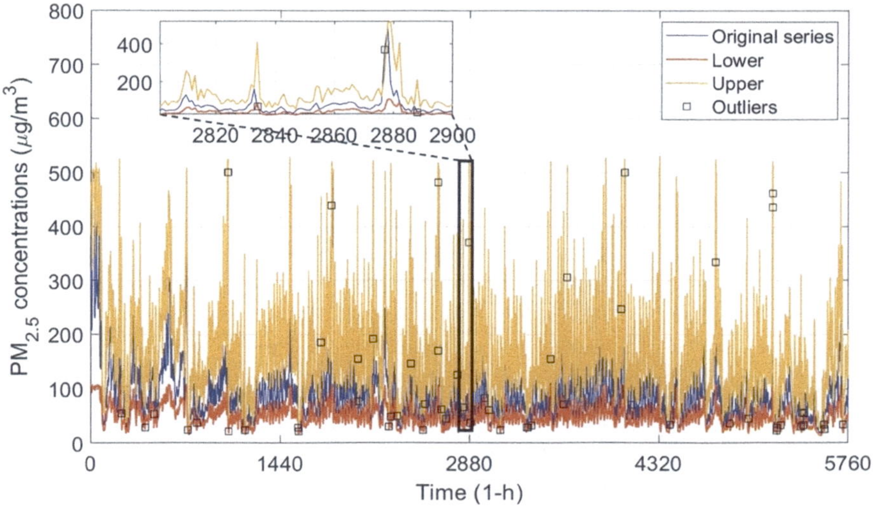

Fig. 2.31 Outlier detection results of the LSTM network

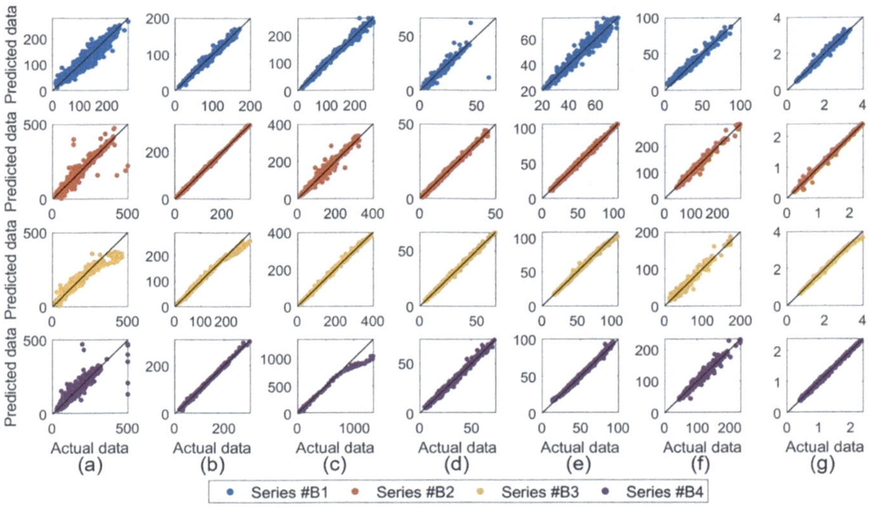

Fig. 2.32 The scatter plots of the predicted data and original data after LSTM detection. (**a**) AQI. (**b**) PM2.5. (**c**) PM10. (**d**) SO_2. (**e**) NO_2. (**f**) O_3. (**g**) CO

1. The KNN method has better missing data imputation performance than the SOM. Taking 25% missed series #A1 as an example, the P values of the KNN and SOM are 0.99 and 0.96 respectively. Although the KNN only uses a single variable, it uses future information to complete missing data. The SOM algorithm only uses multivariate information at the current time. Air pollution data

Table 2.3 The missing data imputation performance of the KNN and SOM for 25% missing ratio

Model	Series	Accuracy performance				Stability performance		
		MAE (μg/m³)	RMSE (μg/m³)	P	KGE	SDE (μg/m³)	DIE (μg/m³)	IWE (μg/m³)
KNN	#A1	1.2424	3.3993	0.9946	0.9895	3.3896	10.6620	5.4879
	#A2	1.1061	2.0229	0.9935	0.9794	2.0201	3.8718	5.1066
	#A3	1.9602	3.3874	0.9983	0.9939	3.3615	5.8508	10.4262
	#A4	1.0878	1.7070	0.9975	0.9924	1.6743	2.6744	4.6381
SOM	#A1	7.3106	16.5415	0.9551	0.9489	16.5792	37.5986	53.5794
	#A2	5.4262	12.0244	0.9595	0.8855	11.8976	26.0869	36.7262
	#A3	5.6406	11.1151	0.9904	0.9751	11.1288	21.9569	30.4478
	#A4	4.1736	9.0787	0.9704	0.9647	9.0861	19.7810	18.1652

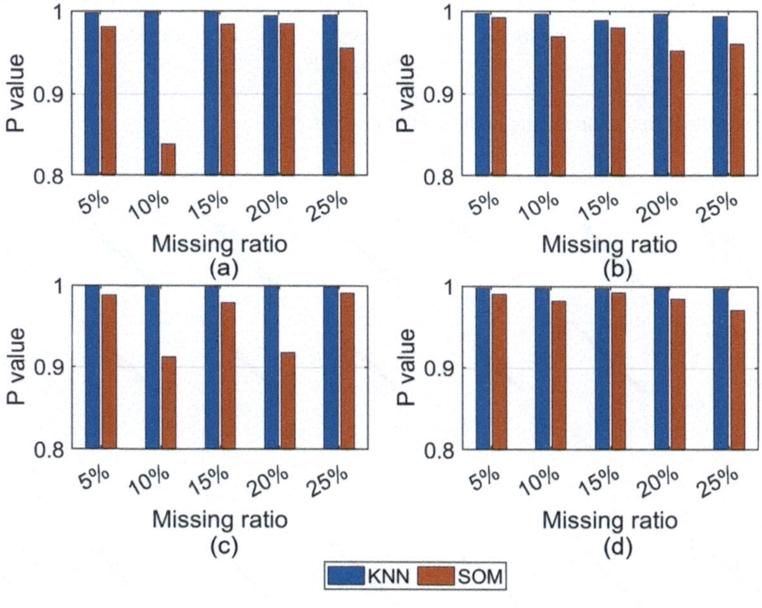

Fig. 2.33 The relation between the imputation performance and missing ratio. (**a**) Series #A1. (**b**) Series #A2. (**c**) Series #A3. (**d**) Series #A4

belongs to time series. The temporal correlation is more important than the correlation cross variables. Therefore, the KNN method performs better than the SOM method. Despite the worse performance, the SOM algorithm has a significant advantage of being able to calculate online because it does not use future information.

2. The missing data imputation performance of the KNN method has little relationship with the missing ratio. Taking series #A1 with 5%, 10%, 15%, 20%, and

25% missing ratios as an example, the P values are 0.9976, 0.9980, 0.9985, 0.9943, and 0.9946 respectively. Because air quality data has a significant time dependence. If the past and future sequences are known, it is easy to estimate missing values. Even if the missing value is relatively large, the KNN method can explore the time series correlation and effectively impute the missing value.
3. The missing data imputation performance of the SOM method has a significant relationship with the missing ratio. Taking series #A2 with 5%, 10%, 15%, 20%, and 25% missing ratios as an example, the P values are 0.9919, 0.9687, 0.9791, 0.9515, and 0.9595 respectively. This is because if the missing ratio is relatively large, there will be more missing variables in the data at the same time. This increases the difficulty of SOM matching templates, thereby reducing the accuracy of missing data imputation.

2.6.2 Performance Comparison of Outlier Detection

In this chapter, three different outlier detection methods are analyzed. These methods have different mechanisms. To ensure comparison fairness, these methods are validated on the same datasets. The performance between the outlier detection methods and without outlier detection is presented in Table 2.4. The comparison of these outlier detection methods is shown in Fig. 2.34.

From Table 2.3 and Fig. 2.34, it can be concluded as follows:

1. The outlier detection method can improve prediction performance. Taking series #B2 as an example, the P values of the isolation forest, Hampel filter, deep learning and non-detection methods are 0.9937, 0.9938, 0.9931 and 0.9910 respectively. This is because the outlier detection method can improve the information quality of the series. The prediction model can better discover the autocorrelation function within the processed series. So, better forecasting performance can be achieved.
2. The Hampel filter method has the best performance in most cases. Taking series #B3 as an example, the P values of the isolation forest, Hampel filter, and deep learning are 0.9933, 0.9960, and 0.9952 respectively. The isolation forest method abandons the time dependence of time series, but detect outliers based on the correlation between multiple variables. In time series, time dependence is stronger than variable dependence. The isolation forest method lacks key information, resulting in insufficient performance. Similar to the Hampel filter, the deep learning method is based on the temporal dependency. However, the deep learning method has large complexity. It will lead to over-fitting and limited performance.

Table 2.4 Performance of the outlier detection methods and without outlier detection

Model	Series	Accuracy performance				Stability performance		
		MAE (μg/m³)	RMSE (μg/m³)	P	KGE	SDE (μg/m³)	DIE (μg/m³)	IWE (μg/m³)
Isolation forest	#B1	2.3827	3.6799	0.9879	0.9603	3.5578	5.7261	9.7546
	#B2	2.4292	4.9469	0.9937	0.9901	4.9403	10.4185	10.6870
	#B3	2.4009	4.8848	0.9933	0.9669	4.8483	9.9885	10.0371
	#B4	3.0996	9.1061	0.9900	0.9624	9.0709	34.8539	11.0702
Hampel filter	#B1	2.6682	3.9679	0.9875	0.9535	3.7294	5.7339	10.7725
	#B2	2.3504	4.7913	0.9938	0.9911	4.7899	10.0453	10.2058
	#B3	2.1003	3.6652	0.9960	0.9874	3.6373	6.3397	8.7851
	#B4	2.5358	5.4698	0.9927	0.9849	5.4566	12.6525	10.2206
Deep learning	#B1	2.6317	3.8740	0.9873	0.9684	3.7804	5.7277	10.7853
	#B2	2.5217	5.1494	0.9931	0.9886	5.1332	10.8870	10.6112
	#B3	2.3639	4.2191	0.9952	0.9781	4.1802	7.4315	10.5023
	#B4	3.0307	8.5016	0.9909	0.9629	8.4661	30.0308	11.6798
Without outlier detection	#B1	2.7551	4.0425	0.9876	0.9500	3.7715	5.5179	10.8466
	#B2	2.6559	5.6952	0.9910	0.9885	5.6941	12.9807	10.9155
	#B3	2.4637	4.5729	0.9946	0.9785	4.5239	8.5777	10.9188
	#B4	3.9183	12.4624	0.9858	0.9414	12.3658	48.7489	12.4624

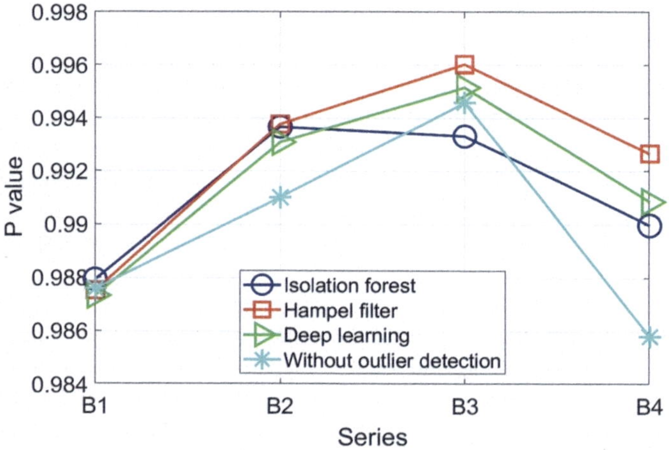

Fig. 2.34 The comparison of these outlier detection methods

2.7 Conclusions

This chapter delves into the critical aspects of data preprocessing for air quality monitoring, encompassing missing data imputation and outlier detection techniques. The study explores both univariate and multivariate methods for addressing missing data, highlighting the effectiveness of the k-nearest neighbors (KNN) method for its simplicity and use of temporal information, while acknowledging the potential of Self-Organizing Maps (SOM) for online imputation though with limited performance. In outlier detection, this chapter evaluates isolation forest, Hampel filter, and a deep learning-based forecasting approach, with the Hampel filter generally demonstrating superior performance due to its strong temporal dependence characteristics and less complex computation. This analysis is based on data from 13 cities in the Jing-Jin-Ji region of China, using 7 air quality variables across 36,000 samples from 2014 to 2018, confirming the temporal and spatial correlation of air pollution in the area, The comparison indicates that effective data preprocessing is essential for enhancing the accuracy and reliability of air quality monitoring and subsequent modeling.

References

Box GEP, Jenkins GM, Reinsel GC, Ljung GM (2015) Time series analysis: forecasting and control. Wiley, Hoboken

Çelik M, Dadaşer-Çelik F, Dokuz AŞ (2011) Anomaly detection in temperature data using dbscan algorithm. In: 2011 international symposium on innovations in intelligent systems and applications. IEEE, pp 91–95

Chai T, Draxler RR (2014) Root mean square error (RMSE) or mean absolute error (MAE)?—Arguments against avoiding RMSE in the literature. Geosci Model Dev 7(3):1247–1250

Gautam C, Ravi V (2015) Data imputation via evolutionary computation, clustering and a neural network. Neurocomputing 156:134–142. https://doi.org/10.1016/j.neucom.2014.12.073

Ghaleb FA, Kamat MB, Salleh M, Rohani MF, Abd Razak S (2018) Two-stage motion artefact reduction algorithm for electrocardiogram using weighted adaptive noise cancelling and recursive Hampel filter. PLoS One 13(11):e0207176

Gneiting T, Balabdaoui F, Raftery AE (2007) Probabilistic forecasts, calibration and sharpness. J R Stat Soc B (Stat Methodol) 69(2):243–268

Gold MS, Bentler PM, Kim KH (2003) A comparison of maximum-likelihood and asymptotically distribution-free methods of treating incomplete nonnormal data. Struct Equ Model 10(1):47–79

Kong Z, Tang B, Deng L, Liu W, Han Y (2020) Condition monitoring of wind turbines based on spatio-temporal fusion of SCADA data by convolutional neural networks and gated recurrent units. Renew Energy 146:760–768. https://doi.org/10.1016/j.renene.2019.07.033

Kornelsen K, Coulibaly P (2014) Comparison of interpolation, statistical, and data-driven methods for imputation of missing values in a distributed soil moisture dataset. J Hydrol Eng 19(1):26–43

LeCun Y, Bengio Y, Hinton G (2015) Deep learning. Nature 521(7553):436

Lin Z, Liu X, Collu M (2020) Wind power prediction based on high-frequency SCADA data along with isolation forest and deep learning neural networks. Int J Electr Power Energy Syst 118:105835. https://doi.org/10.1016/j.ijepes.2020.105835

Liu H, Shah S, Jiang W (2004) On-line outlier detection and data cleaning. Comput Chem Eng 28(9):1635–1647. https://doi.org/10.1016/j.compchemeng.2004.01.009

Liu FT, Ting KM, Zhou ZH (2008) Isolation forest. In: 2008 eighth IEEE international conference on data mining. IEEE, pp 413–422

Liu H, Xu Y, Chen C (2019) Improved pollution forecasting hybrid algorithms based on the ensemble method. Appl Math Model 73:473–486. https://doi.org/10.1016/j.apm.2019.04.032

Qin Y, Lou Y (2019) Hydrological time series anomaly pattern detection based on isolation forest. In: 2019 IEEE 3rd information technology, networking, electronic and automation control conference (ITNEC). IEEE, pp 1706–1710

Sharadga H, Hajimirza S, Balog RS (2020) Time series forecasting of solar power generation for large-scale photovoltaic plants. Renew Energy 150:797–807. https://doi.org/10.1016/j.renene.2019.12.131

Ting JA, Theodorou E, Schaal S (2007) A Kalman filter for robust outlier detection. In: 2007 IEEE/RSJ international conference on intelligent robots and systems. IEEE, pp 1514–1519

Tutz G, Ramzan S (2015) Improved methods for the imputation of missing data by nearest neighbor methods. Comput Stat Data Anal 90:84–99. https://doi.org/10.1016/j.csda.2015.04.009

Zhang Y, Zheng H, Liu J, Zhao J, Sun P (2018) An anomaly identification model for wind turbine state parameters. J Clean Prod 195:1214–1227. https://doi.org/10.1016/j.jclepro.2018.05.126

Chapter 3
Data Decomposition in Air Quality Monitoring

Abstract This chapter explores the application of data decomposition techniques in air quality monitoring, focusing on wavelet decomposition and modal decomposition methods. These advanced techniques are particularly effective for analyzing the nonlinear, non-stationary, and multi-scale characteristics inherent in air quality data. By employing multi-scale decomposition, researchers can extract valuable insights, including long-term trends, seasonal variations, and random fluctuations. Such detailed analysis not only enhances the accuracy of air quality assessments but also provides robust data support for policy formulation, environmental management, and pollution control strategies. These methods play a crucial role in addressing complex environmental challenges.

3.1 Introduction

Air quality is closely connected to public health, environmental sustainability, and the social economy's development. With rapid industrial growth and urban expansion, pollution levels are escalating, becoming a critical issue globally. The World Health Organization highlights that air pollution is responsible for millions of premature deaths annually, posing a serious threat to public health.

To effectively monitor and improve air quality, accurate analysis of air quality data is particularly important. However, air quality assessments require understanding atmospheric pollution and its evolving trends, offering strong data and theoretical foundations for shaping environmental policies. Typically, air quality data is represented as multivariate time series, which presents challenges such as large datasets, high dimensionality, and insufficient labeled information (Luo et al. 2024). Fujiwara et al. (2021) explore methods for analyzing the complexity of air quality data, noting that traditional linear models struggle to capture the multidimensional characteristics and nonlinear relationships inherent in this data. This limitation impedes their ability to accurately forecast air quality patterns. Moreover, air quality is influenced by a range of non-linear and multiscale factors, including

H. Liu et al., *Data Science in Air Quality Monitoring*, Engineering Applications of Computational Methods 23, https://doi.org/10.1007/978-981-96-5777-3_3

meteorological conditions, geographical variables, and human activities, which complicates the use of linear analysis methods.

In recent years, data decomposition technology has received widespread attention in processing complex environmental data. Among them, wavelet decomposition and modal decomposition methods have become research hotspots due to their advantages in nonlinear and non-stationary signal analysis. Sifuzzaman et al. (2009) noted that wavelet transforms are often more reliable and effective than Fourier transforms in signal processing. Silik et al. (2021) said that wavelet transform has considerable potential in digital signal processing Klionskiy et al. (2017) demonstrated that EMD (Empirical Mode Decomposition) has been effectively used for signal denoising, particularly in cases involving homoscedastic and heteroscedastic noise.

3.1.1 Application of Wavelet Decomposition in Air Quality Data Analysis

The analysis of non-stationary time series (TS) has gained substantial interest in recent decades across various scientific disciplines. Decomposition methods were initially developed to extract distinct components such as trends, seasonal patterns, and abrupt changes, which are prevalent in the temporal variability of time series (Rhif et al. 2019). Recent studies have shown that wavelet analysis, when combined with artificial neural networks (ANN), provides enhanced predictive capabilities. For instance, research by Wang et al. demonstrated that this hybrid model predicts PM2.5 concentrations with greater accuracy (Guo et al. 2023b). Further, wavelet decomposition has proven effective in identifying meteorological factors that influence pollutant levels, thus improving air quality forecasting (Guo et al. 2023a). Wavelet decomposition is a powerful time-frequency analysis tool that can perform multi-scale analysis of signals in both time and frequency domains. It can effectively extract local features and mutation information in signals by selecting appropriate wavelet basis functions and decomposition layers. Wavelet decomposition has been widely used in the analysis of air quality data. For example, researchers used wavelet decomposition to process PM2.5 concentration data and successfully extracted its long-term trend and seasonal variation characteristics, which helps to understand the source and propagation of pollutants.

3.1.2 Application of Modal Decomposition in Air Quality Data Analysis

The modal decomposition method, particularly Empirical Mode Decomposition (EMD), is a data-driven, adaptive technique that does not require predefined basis functions. Instead, EMD adapts to the signal, breaking it down into several intrinsic mode functions (IMFs) that reflect the data's inherent characteristics. This method has unique advantages in processing nonlinear and non-stationary data. In air quality data analysis, EMD is used to extract periodic components and random fluctuations in pollutant concentrations to help identify potential influencing factors and pollution events. Niu et al. (2022) introduced an enhanced EMD technique to decompose air pollution data and predict pollutant concentrations through a deep learning model. This approach is particularly suitable for handling nonlinear and highly volatile data.

3.1.3 Deficiencies and Challenges of Existing Research

Although wavelet decomposition and modal decomposition have achieved certain results in air quality data analysis, there are still some problems that need further study. For example, wavelet decomposition requires the selection of appropriate wavelet basis functions and is sensitive to noise; the EMD method may be affected by endpoint effects and modal aliasing, affecting the stability and accuracy of the decomposition results. In addition, there are differences in accuracy and efficiency between the two methods to separate high-frequency and low-frequency signals.

When comparing modal decomposition (such as EMD, EEMD, CEEMD) and wavelet decomposition, we can analyze the differences, advantages and disadvantages, and applicable scenarios of the two from multiple dimensions. The following is a detailed comparison:

3.1.4 Temporal Resolution

The time resolution of modal decomposition is linked to the local characteristics of the signal, and it is not constrained by a fixed resolution. Wavelet decomposition offers good time resolution, particularly for high-frequency components, making it effective for detecting rapid variations. Wavelet decomposition has a high time resolution on high-frequency components and is suitable for analyzing short-term emergencies. Modal decomposition adaptively reflects the characteristic changes of the signal and has a more flexible time resolution.

3.1.5 Frequency Resolution

Modal decomposition adaptively extracts different frequency components from the signal, and does not rely on the preset basis function. As a result, the frequency resolution is influenced by the signal's inherent characteristics and reflects its natural frequency more effectively.

The frequency resolution of wavelet decomposition depends on the selected wavelet basis and the number of decomposition levels. It provides good resolution for low-frequency components, while high-frequency resolution may be limited. The high-frequency resolution is poor, while the low-frequency resolution is good, which is determined by the multi-scale characteristics of wavelet decomposition.

3.1.6 Boundary Effect

EMD is prone to boundary effect, which causes the decomposition result to fluctuate and be unstable at the signal boundary. EEMD and CEEMD improve the boundary effect by adding noise. VMD further reduces the boundary effect through frequency domain optimization. Wavelet decomposition also has boundary effect problem, but the boundary distortion can be partially alleviated by filling the signal boundary.

3.1.7 Noise Reduction Effect

Modal decomposition can better separate the main components of noise and signal. For example, high-frequency IMF can be regarded as noise removal to achieve noise reduction. EEMD and CEEMD perform better in noise reduction.

Wavelet decomposition removes high-frequency noise components by setting a threshold to achieve noise reduction effect. It is particularly suitable for filtering in the high-frequency band of noise signals.

- Modal decomposition and wavelet decomposition each have their unique strengths in signal processing. Modal decomposition is particularly well-suited for handling nonlinear and non-stationary signals, offering high adaptability in extracting the intrinsic modes of the signal. On the other hand, wavelet decomposition excels in computational efficiency and is highly effective for multi-scale analysis of local stationary signals, such as in image processing and signal denoising. Depending on the type of signal and the specific analysis requirements, the appropriate method can be selected to optimize the decomposition process.

To further study the application of data decomposition technology in air quality monitoring, this study needs to obtain high-quality air quality data. The data acquisition process includes steps such as determining the data source, selecting measurement parameters, formulating data collection methods, and data preprocessing to ensure that the data used is reliable and representative.

As the world's second-largest economy, China faces significant environmental pollution challenges, particularly concerning ambient air pollution, which has become a major public health risk (Yang et al. 2017). Beijing, as the capital of China, has made remarkable achievements in improving air quality in recent years. Improvements in air quality are crucial to residents' health, as long-term exposure

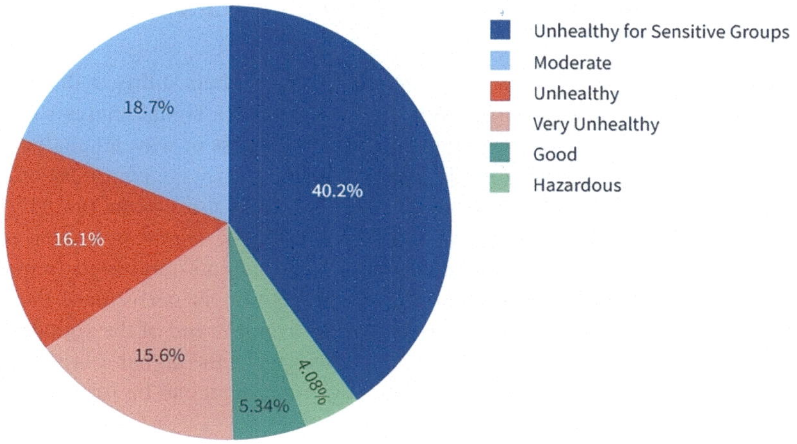

Fig. 3.1 Air quality distribution pie chart over 6 years

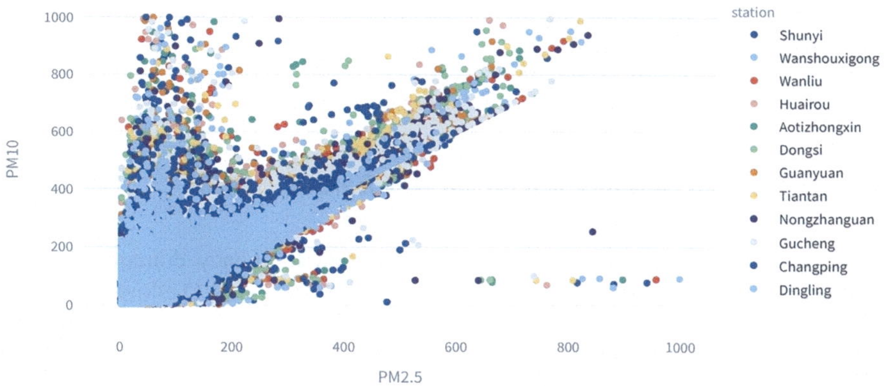

Fig. 3.2 The relative relationship between PM10 and PM2.5

to polluted air increases the risk of cardiovascular and respiratory diseases. Fine particulate matter (PM2.5), with a diameter of 2.5 micrometers or less, is strongly linked to premature mortality and can travel long distances, impacting air quality and public health on a regional and even intercontinental scale (Anenberg et al. 2014). Therefore, the Beijing municipal government has taken measures such as restricting motor vehicles, promoting clean energy, improving public transportation, and controlling industrial emissions to reduce pollutant emissions. The air quality data used in this study are related to air quality data from 12 stations in Beijing from 2013 to 2017. The distribution of air quality categories, air pollutant concentrations (PM2.5, sulfur dioxide, CO), and the correlation between air pollutant concentrations are visualized.

The air quality can be divided into six levels. The distribution of air quality at 12 stations over 6 years is shown in Fig. 3.1:

The correlation between PM2.5 and PM10 at 12 stations is shown in Fig. 3.2:

3.2 Wavelet Decomposition of Air Quality Data

Wavelets are an effective tool for signal denoising due to their ability to decompose a signal into different scales, which significantly improves signal analysis (Cohen 2012). Wavelet Decomposition (WD) is the primary process of wavelet analysis and has been widely used in various fields, including data and image compression, partial differential equation solving, transient detection, and noise reduction (Dautov and Özerdem 2018). By applying wavelet transform, the signal is broken down layer by layer into low-frequency (approximate coefficients) and high-frequency (detail coefficients) parts. Each time the decomposition is performed, the low-frequency part retains the main characteristics and overall trend of the signal, while the high-frequency part retains the local detail changes of the signal. For the multi-layer decomposition process, the low-frequency component can be further decomposed to analyze the signal at different scales.

The basic principle of WD is to decompose the complex signal into approximation and detail parts at different scales by multi-scale decomposition of the signal. Specifically, wavelet transform uses a set of basic functions generated by the mother wavelet to project the signal and obtain wavelet coefficients of different scales and positions. These coefficients reflect the characteristics of the signal at different scales. Here are some of its features:

3.2.1 Time-Frequency Localization Characteristics

WD provides good resolution in both the time and frequency domains. Unlike Fourier transform, which only analyzes the frequency domain, WD allows for localized analysis in both time and frequency. This time-frequency localization means WD can adaptively adjust the resolution of time and frequency based on the signal scale, making it ideal for analyzing short-term high-frequency signals or long-term low-frequency ones. As a result, the local details of the signal can be effectively captured.

3.2.2 Multi-resolution Analysis

Wavelet transform has the ability of multi-resolution analysis, which means that it can decompose the signal at different scales from coarse to fine. Wavelet transform analyzes the details of each level of the signal step by step by decomposing the high-frequency and low-frequency parts of the signal at different scales. This multi-resolution feature is very suitable for processing signals with multi-scale characteristics, such as edge detection or image compression in image processing.

3.2.3 Strong Sparse Representation Capability

WD can provide sparse representation capability in many signal processing tasks. For most signals, the coefficients after WD are concentrated on a few high-energy coefficients, while other coefficients are close to zero. This makes WD very effective in applications such as signal compression and noise reduction and can significantly reduce the amount of data while retaining the main features of the signal. It is precisely because of this sparse representation feature that WD is widely used in image compression (such as JPEG2000) and other data processing fields.

Wavelet decomposition is a technique for decomposing a signal into components of different scales and frequencies and is often used in signal processing and time series analysis. Depending on different needs, common wavelet decomposition types include discrete wavelet transform (DWT) and continuous wavelet transform (CWT).

The continuous wavelet transform (CWT) is often used to generate spectrograms that illustrate the frequency content of sounds (or other signals) over time, in a way that is similar to how music is analyzed (Lang and Forinash 1998).

DWT is more suitable for PM2.5 time series data. These transforms can decompose the data at multiple scales and help identify trends, seasonal components, and noise components at different scales. In addition, commonly used wavelet bases (such as Daubechies, Symlets, Coiflets, and Biorthogonal) can be applied to DWT, and the appropriate wavelet base can be selected based on the data characteristics.

- DWT and CWT (also known as WT) each have distinct characteristics when applied to PM2.5 data decomposition. The choice of method depends on the specific analysis goals and the data's properties. Below are key aspects of DWT and CWT in PM2.5 data analysis: Time resolution: DWT is a discrete transformation that breaks the signal into multiple levels of low-frequency (approximate coefficients) and high-frequency (detail coefficients) parts, using binary scaling and translation. It has a fixed time-frequency resolution: the low-frequency components offer higher frequency resolution, while the high-frequency components provide better time resolution. CWT, in contrast, continuously decomposes the signal into different frequency and time scales by translating the changing wavelet basis functions. This allows for a finer time-frequency resolution that can be adjusted more flexibly, especially for analyzing high- and low-frequency data.
- Computational complexity: DWT is computationally efficient because it only down samples and decomposes the low-frequency components of each layer. This makes DWT computationally smaller and occupies less storage resources, making it suitable for processing large amounts of data. CWT has high computational complexity because CWT performs convolution operations on the signal at each scale and does not down sample, thus retaining more information. This results in large amounts of data, long computation times, and higher storage requirements.
- Time-frequency localization: DWT has multi-resolution analysis characteristics and can capture trends and details of signals at different scales. DWT is suitable

for layered observation of long-term and short-term trends in PM2.5 data through binary scaling and translation of wavelet basis functions. CWT has good time-frequency localization characteristics and can observe short-term changes in PM2.5 data at higher frequencies and capture long-term trends at lower frequencies. CWT can generate time-frequency diagrams to show the time-frequency evolution of PM2.5 concentrations, which is suitable for observing dynamic changes.

- Data sampling and time-shift invariance: DWT decomposes the signal layer by layer through down sampling, which may introduce some time-shift deviations during the decomposition process, thus affecting the time resolution. CWT does not perform down sampling, so it maintains time-shift invariance, that is, the translation of the signal does not change the decomposition result. This is particularly important for analyzing sudden events and abnormal changes at specific moments in the PM2.5 time series.
- Processing results: The result of DWT decomposition is a series of multi-scale coefficients, which can analyze the long-term trend and high-frequency noise of the signal at each level. The layered characteristics of DWT allow us to observe the performance of PM2.5 data at different frequencies layer by layer, which is suitable for denoising or multi-scale analysis. The result of CWT decomposition is a time-frequency diagram that can intuitively show the frequency changes of the signal over time. CWT can accurately locate the time and frequency components of PM2.5 concentration changes and is suitable for analyzing non-stationary and dynamically changing pollution data.

DWT is more suitable for scenarios where PM2.5 data needs to be processed quickly and analyzed in simple layers, such as trend analysis and noise reduction. It is computationally efficient, but compromises in time and frequency resolution, and is suitable for applications that require long-term monitoring and trend extraction. CWT is suitable for in-depth analysis of the time-frequency characteristics of PM2.5 concentrations and is more effective in the dynamic analysis of non-stationary signals, but has high computational complexity. CWT can generate time-frequency graphs, which are very suitable for more complex situations that require detailed observation of pollution events, frequency changes, etc.

3.2.4 Discrete Wavelet Transform

Discrete Wavelet Transform is a technique for multi-resolution analysis of signals. DWT decomposes a signal into frequency components of different scales to analyze its characteristics at different scales. The core idea of this method is to decompose the signal by selecting different wavelet basis functions and using these basis functions to capture the local characteristics of the signal (Brown Ingram 2009). The following is a detailed introduction to DWT, including basic principles, decomposition and reconstruction process, mathematical description and application.

Discrete wavelet transform realizes multi-resolution analysis (MRA) of signals by decomposing them at different scales. The characteristic of this method is that the resolution of the signal in the low-frequency part is gradually increased to analyze the trend more accurately; while in the high-frequency part, the resolution is gradually reduced to highlight the detailed features. DWT realizes this multi-resolution decomposition through filtering and down sampling operations.

Low-pass filter extracts the low-frequency components of the signal to reflect the overall trend of the signal.

High-pass filter extracts the high-frequency components of the signal to reflect the details and rapid changes of the signal.

Down sampling: Down sampling refers to reducing the sampling frequency or the number of data, for example, reducing daily data to weekly data or extracting low-frequency signals from high-frequency signals. In time series data, down sampling is often used to reduce data density to simplify analysis and reduce computational costs. It can also help remove high-frequency noise and make trends more obvious. In the processing of unbalanced data sets, down sampling reduces the number of samples in the majority class to make the number of samples in different classes more balanced. The specific operation of down sampling is to perform a sampling operation on the filtered signal sequence with a sampling interval of 2, that is, only retaining samples at even positions. This can effectively reduce the amount of data and improve computational efficiency.

Through low-pass and high-pass filtering followed by down-sampling, the signal is separated into two components: low-frequency (approximation) and high-frequency (details). This process can then be repeated on the low-frequency component to break it down into more refined low-frequency and high-frequency components. This technique is known as multi-layer wavelet decomposition. Let the original signal be $x[n]$, the low-pass filter be $h[n]$, and the high-pass filter be $g[n]$. Through filtering and down sampling, we can get the low-frequency and high-frequency decomposition coefficients:

Low frequency (approximation) coefficients:

$$a_j[k] = \sum_n x[n] \cdot h[n-2k] \tag{3.1}$$

High frequency (detail) factor:

$$d_j[k] = \sum_n x[n] \cdot g[n-2k] \tag{3.2}$$

The subscript j here indicates the scale (or level) of decomposition. Usually, we will recursively decompose the low-frequency part from the first level until the preset number of decomposition levels is reached.

Wavelet and scaling function:

- Scaling function $\phi(t)$: The scaling function is used to represent the low-frequency part of the signal. By recursively scaling and moving the scaling c

- Wavelet function $\psi(t)$: The wavelet function is used to represent the high-frequency part of the signal, mainly used to analyze the details and mutation points of the signal.

The scaling function and wavelet function can be defined by the following dual scaling equation:

$$\phi(t) = \sum_{n} h[n] \cdot \phi(2t - n) \tag{3.3}$$

$$\psi(t) = \sum_{n} g[n] \cdot \phi(2t - n) \tag{3.4}$$

These equations define the recursive relationship between the scaling function and the wavelet function through the filter coefficients.

Taking a simple one-dimensional signal as an example, assume that the signal $x[n]$ is decomposed by DWT as follows:

First-level decomposition: Low-pass and high-pass filters are performed on the signal $x[n]$ to obtain the first-level low-frequency coefficient a_1 and high-frequency coefficient d_1 respectively.

Second-level decomposition: Low-pass and high-pass filters are performed on the first-level low-frequency coefficient a1 to obtain the second-level low-frequency coefficient a_2 and high-frequency coefficient d_2.

Recursive decomposition: The low-frequency coefficients can be further decomposed at a higher level to obtain finer-grained signal components.

This decomposition process will continue until the preset decomposition level j is reached, and low-frequency and high-frequency information at each level is obtained. This is the core of multi-scale analysis. This is central to multi-scale analysis, where the low-frequency part captures the overall trend of the signal through the scaling function, and the high-frequency part detects the detailed changes using the wavelet function.

The signal reconstruction process (inverse discrete wavelet transform, IDWT) aims to restore the low-frequency and high-frequency coefficients of each level to reconstruct the original signal. The reconstruction process involves the following steps: Up-sampling: Unsampled the low-frequency and high-frequency coefficients of each layer, that is, insert a zero value between each sampling point to restore the sampling rate of the signal.

Convolution: Low-pass and high-pass filter convolutions are applied to the up-sampled low-frequency and high-frequency coefficients to restore the corresponding components of the signal.

Superposition: Superimpose low-frequency and high-frequency signals of different scales to obtain a reconstructed signal. Since the filter bank meets the perfect reconstruction condition, the original signal can be reconstructed without distortion.

IDWT is the inverse operation of discrete wavelet transform (DWT), which recovers the signal by reconstructing its low-frequency and high-frequency components. In the DWT process, the signal is decomposed into low-frequency and high-frequency components at multiple resolution levels. IDWT combines these components step by step to restore the signal. It uses a set of inverse filters to restore the signal through interpolation, convolution, and superposition of the low-frequency and high-frequency coefficients of each decomposition layer. The inverse filter in the IDWT process corresponds to the inverse of the DWT filter, which is applied to the low-frequency and high-frequency components using low-pass and high-pass inverse filters, respectively. The low-pass inverse filter restores the overall trend of the signal, while the high-pass inverse filter recovers the details. IDWT can merge the details and trend information decomposed into each layer and finally restore the signal close to the original state, which makes it crucial in multi-resolution analysis. In general, IDWT, as the inverse process of DWT, is an indispensable part of wavelet analysis. Its role is to reconstruct the multi-scale, wavelet-decomposed signal to its original form, making the signal feature analysis after wavelet decomposition more practical.

We visualized the PM2.5 data of the Temple of Heaven subway station and applied different wavelet functions to perform multi-level decomposition. The reconstructed signal results after three, four, and five different levels of data decomposition are shown in Figs. 3.3, 3.4, and 3.5.

Taking the five-level decomposition as an example, we introduce the signal decomposition results after multi-level decomposition using Daubechies 4 (db4) wavelet. The specific decomposition is as follows:

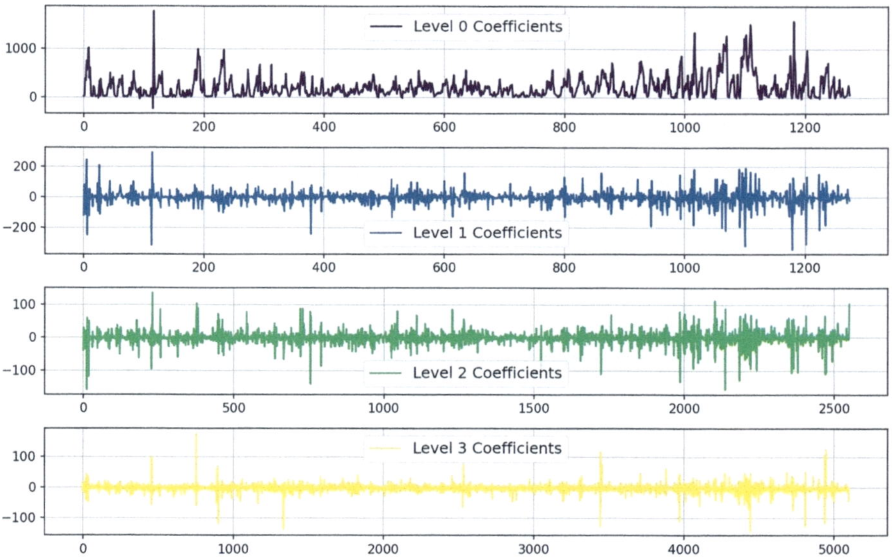

Fig. 3.3 db4 wavelet decomposition—level 3

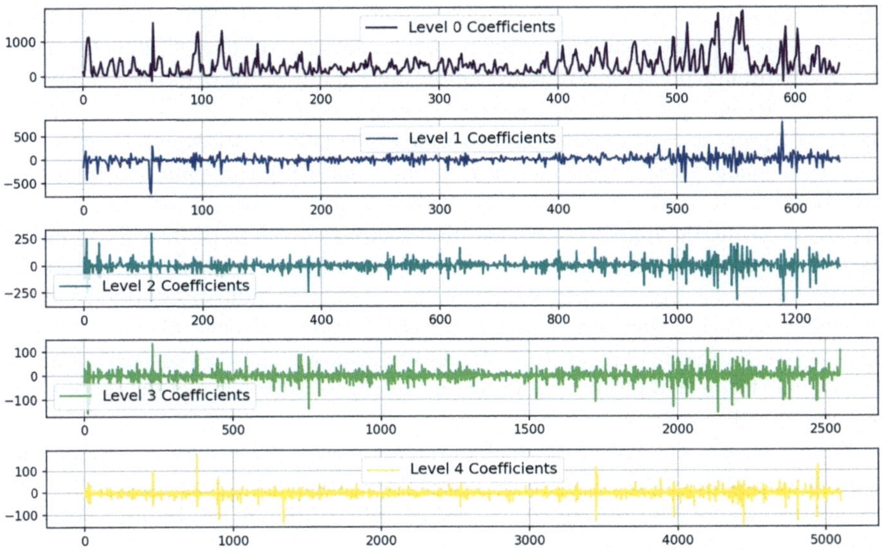

Fig. 3.4 db4 wavelet decomposition—level 4

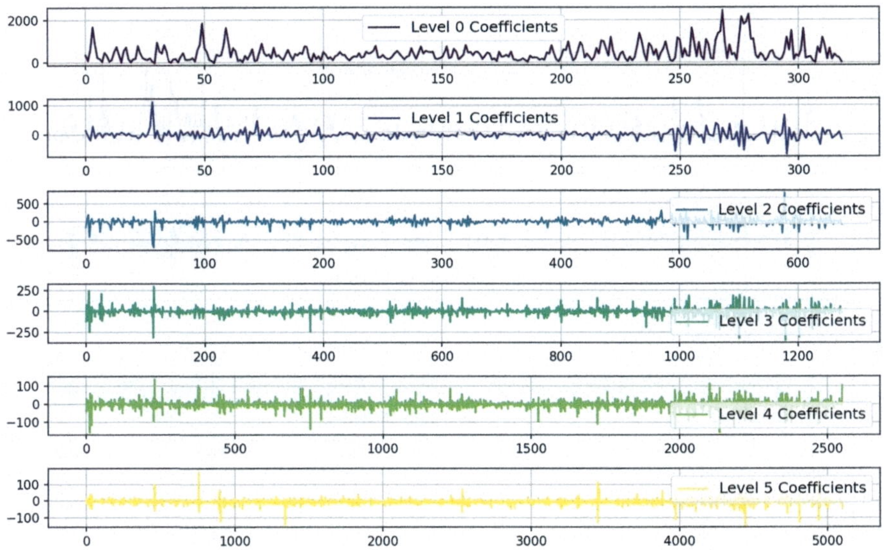

Fig. 3.5 db4 wavelet decomposition—level 5

3.3 Top Layer: Approximation Coefficients

This blue curve represents the low-frequency component of the signal at level 5 (i.e., the overall trend). The signal has a relatively smooth change at low frequencies, reflecting the overall shape and trend of the signal.

3.4 Detail Coefficients

Level 1 detail coefficients (green): represent the detail changes of the highest frequency component, capturing the finest features of the signal. This layer usually contains many fast fluctuations, reflecting short-term changes.

Level 2 detail coefficients (red): As the decomposition level decreases, this layer reflects lower frequencies and a smaller fluctuation amplitude, but can still capture some relatively fast changes.

Level 3 detail coefficients (cyan): At this layer, the mid-frequency components in the signal are retained, and the fluctuations are smoother, mainly reflecting changes in lower frequencies.

Level 4 detail coefficients (purple): lower frequency detail changes. Compared with the previous layers, the fluctuations become gentler, focusing on longer-term fluctuations in the signal.

Level 5 detail coefficients (yellow): the lowest frequency detail changes, close to the trend component of the signal, with almost no obvious high-frequency fluctuations.

In general, as the number of decomposition levels increases, the frequency range of the detail coefficients decreases, and the fluctuations become smoother. The approximate coefficients at higher levels, such as the fifth level, preserve the overall trend of the signal, while the detail coefficients at each level capture finer high- and medium-frequency details. This multi-level decomposition approach allows us to observe the characteristics of the signal at various scales, thus enabling comprehensive analysis in the time-frequency domain. Comparing the results of different levels of multi-level wavelet decomposition (levels 3, 4, and 5), the following conclusions can be drawn: As the level of decomposition increases (from level 3 to level 5), the waveform of the low-frequency component becomes smoother.

At higher decomposition levels, trend information over longer time frames is retained and the frequency of fluctuations is further reduced. The level 3 approximation coefficient contains more high-frequency components, making its fluctuations more intensive. At level 5, the fluctuation frequency decreases, and the curve is closer to the overall trend of the signal. The detail coefficients are decomposed step by step from level 1 to level 5, and high-frequency components are gradually reduced.

With each additional level of decomposition, less detailed information is retained and becomes smoother. The first level detail coefficient changes fastest among all levels of decomposition, has a larger amplitude, and contains many high-frequency components; while at the fifth level, the detail coefficient amplitude is significantly reduced, the frequency is lower, and mainly retains lower-frequency fluctuations information. In lower levels of decomposition (such as level 3), the frequencies of each level are higher and mainly capture short-term detailed information in the signal. As the decomposition level increases (such as level 5), lower frequency information is retained, which mainly reflects the long-term trend of the signal rather than the details. Therefore, the higher the level of decomposition, the smoother the signal fluctuations become.

In short, lower-level decomposition (such as level 3) is more suitable for analyzing high-frequency detailed information in the signal, while higher-level decomposition (such as level 5) helps capture the overall trend and low-frequency components of the signal. Through multi-level decomposition, different frequency components of the signal can be analyzed more comprehensively. The decomposition results from levels 3 to 5 show the process of signal denoising and smoothing step by step, which is suitable for different types of signal processing and analysis needs.

After performing wavelet decomposition, it is very important to evaluate the quality of the decomposition. Usually, the effect of the decomposition is evaluated through quantitative analysis to determine whether the wavelet decomposition can effectively capture the main features and details of the signal. The following are some common quantitative evaluation methods for judging the quality and effect of wavelet decomposition:

3.4.1 Reconstruction Error

Reconstruction error is a key indicator for evaluating the accuracy of wavelet decomposition and reconstruction. By comparing the difference between the original signal and the reconstructed signal, the effect of decomposition can be judged. Commonly used error indicators include Mean Squared Error (MSE), Root Mean Squared Error (RMSE) and Mean Absolute Error (MAE).

3.4.2 Signal-to-Noise Ratio (SNR)

The signal-to-noise ratio is used to evaluate the fidelity of the signal, usually expressed in decibels (dB). The higher the SNR, the more effective signal components are retained by the wavelet decomposition.

$$\mathrm{SNR} = 10\log_{10}\left(\frac{\sum_{i=1}^{N} x_i^2}{\sum_{i=1}^{N}\left(x_i - \hat{x}_i\right)^2}\right)\frac{n!}{r!(n-r)!} \tag{3.5}$$

If the decomposition and reconstruction quality is good, the SNR should be high.

3.4.3 Correlation Coefficient

The correlation coefficient *r* measures the similarity between the original signal and the reconstructed signal. Its value ranges from −1 to 1, and a value close to 1 indicates that the similarity between the decomposed and reconstructed signals is very high.

By combining these quantitative indicators, we can better judge the effect of wavelet decomposition and ensure that the decomposition can effectively capture the signal characteristics. The reconstruction error (MSE), signal-to-noise ratio (SNR), and correlation coefficient of different decomposition levels are calculated to more comprehensively evaluate the effect of wavelet decomposition.

Level 3: MSE = 1.4087048038904179e-27, SNR = 309.55 dB, Correlation Coefficient = 1.0000.

Level 4: MSE = 1.7922957082835778e-27, SNR = 308.50 dB, Correlation Coefficient = 1.0000.

Level 5: MSE = 2.303921947175242e-27, SNR = 307.41 dB, Correlation Coefficient = 1.0000.

The MSE is small, the SNR is high, and the correlation coefficient is 1. These results show that the fidelity of wavelet decomposition and reconstruction to the original signal is very high, and the reconstructed signal is almost exactly the same as the original signal.

With the increase of decomposition level, MSE increases slightly, and SNR decreases slightly, but these changes are very small, indicating that even if the decomposition level increases, the reconstruction effect remains very good.

From these evaluation indicators, the effect of increasing the decomposition level on the reconstruction quality is negligible, indicating that the wavelet basis function (db4) used can capture the characteristics of the signal well.

Theoretically, under ideal conditions, the data after wavelet decomposition should have no error with the original signal during reconstruction.

In actual situations, due to boundary effects, floating point precision limitations, and error accumulation during multi-layer decomposition, there will be certain errors after reconstruction.

When there are more decomposition levels, the error tends to increase, because the more decomposition levels, the accumulation of boundary effects, the loss of detail information, and the loss of calculation accuracy will increase.

Therefore, in actual use, we need to find a balance in the choice of decomposition levels: effectively extracting the characteristic information of the signal without introducing too much error due to too many decomposition levels. Generally speaking, it is reasonable to choose an appropriate decomposition level (for example, not more than $\log_2(N)$ to avoid excessive error accumulation.

3.4.4 Various Wavelet Basis Functions

Wavelet basis functions are regarded as the core of wavelet transform, which determines how the signal can be effectively decomposed at different scales. Wavelet basis functions provide multi-resolution analysis capabilities, allowing signals to be flexibly converted between time domain and frequency domain, thereby capturing the details and global characteristics of the signal at different levels (Vetterli 1995). For the analysis of PM2.5 concentration time series data, it is crucial to select a suitable wavelet basis function. PM2.5 data usually contains long-term trends and short-term fluctuations, as well as some sudden peaks and outliers. To this end, we need a wavelet basis function that can effectively capture the local details and overall trends of the signal in the time and frequency domains.

3.4.4.1 Daubechies Wavelet (dbN)

Daubechies constructed a compactly supported orthogonal wavelet basis through a recursive algorithm. These wavelet bases are composed of low-pass filters and high-pass filters, have good resolution in the frequency domain, and can efficiently decompose the various frequency bands of the signal (Daubechies 1988). In previous experiments, we used Daubechies wavelet as the wavelet basis function, and the experimental results showed that we achieved a good decomposition effect. Research shows that Daubechies wavelet has a higher classification accuracy while maintaining data features, but due to its longer support interval and complex computational requirements, the required computation time is relatively long. Haar wavelet has higher computational efficiency, but is slightly inferior in terms of feature retention (Sharif and Khare 2014).

The Daubechies wavelet has tight support and smoothness and is suitable for capturing the non-stationary characteristics of time series signals. It can balance resolution in time and frequency and is suitable for multi-level decomposition of signals. PM2.5 data contains long-term trends and some short-term fluctuations, and dbN wavelets (such as db4) can well separate these components of different scales. The db4 wavelet has 4 coefficients and can well balance time and frequency resolution and is a commonly used wavelet basis. The Daubechies wavelet is suitable for daily air quality monitoring, trend detection, and abnormal peak analysis, such as peaks within a day, weekend effects, or air quality changes on holidays.

3.4.4.2 Symlets Wavelet

Symlets is a wavelet basis that balances symmetry and mathematical precision, and is particularly suitable for tasks that are sensitive to signal edges or require high-precision reconstruction (Chavan et al. 2011). Symlets wavelet is a symmetric improvement of Daubechies wavelet. It further improves symmetry and reduces phase distortion in signal reconstruction while maintaining tight support and smoothness. PM2.5 data may contain many sudden pollution events. Symlets wavelet can reduce phase distortion, which is especially helpful for signal denoising and feature extraction. Commonly used sym4 and sym8 can effectively capture short-term pollution fluctuations. Symlets wavelet is suitable for air quality prediction and analysis, especially in cases where accurate restoration of signal shape is required, such as detecting pollution sources, analyzing short-term anomalies and emergencies, etc. We use Symlets wavelet to perform multi-level decomposition of PM2.5 time series data, and the results are shown in Fig. 3.6.

3.4.4.3 Coiflets Wavelet (coifN)

Wei et al. (1997) generalizes an existing family of wavelets, coiflets, by replacing the zero-centered vanishing moment condition on scaling functions with a nonzero-centered one. This modification results in a novel class of compactly supported orthonormal wavelets, referred to as generalized coiflets. This generalization introduces an additional free parameter—the center of mass of the scaling function—which can be adjusted to enhance the characteristics of the resulting wavelet system. These improvements include near-symmetry of the scaling functions and wavelets, near-linear phase of the filter banks, and better sampling approximation properties. Consequently, these new wavelets show promise for a wide range of applications in signal processing and numerical analysis.

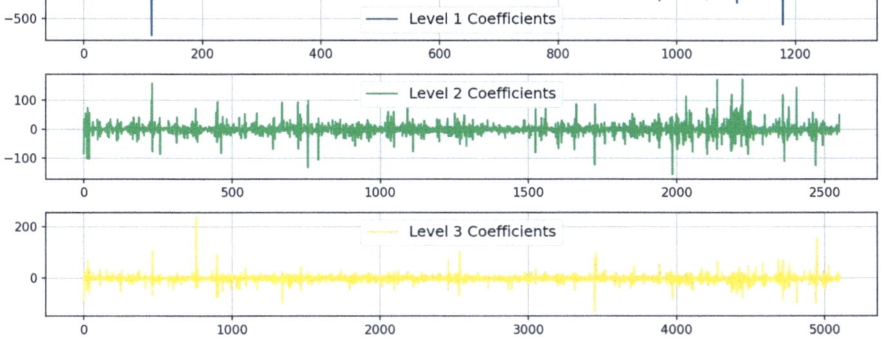

Fig. 3.6 sym4 wavelet decomposition—level 3

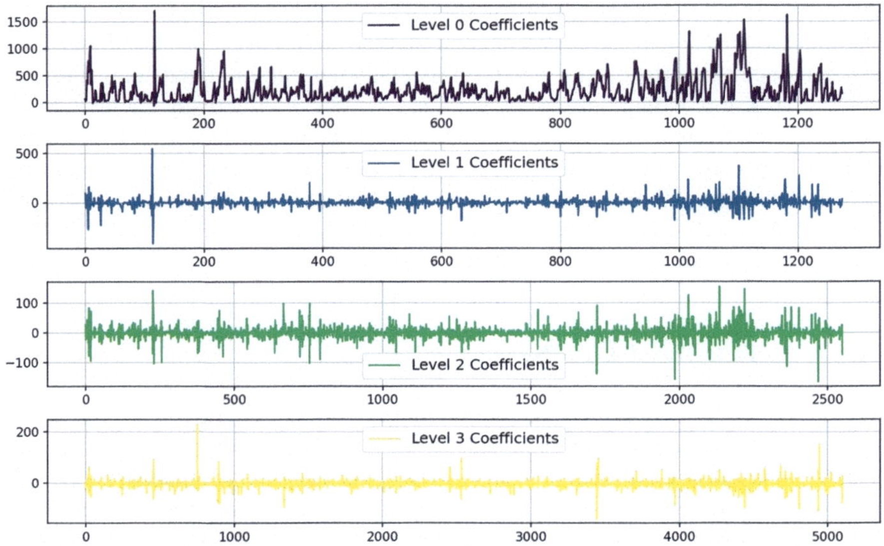

Fig. 3.7 coif4 wavelet decomposition—level 3

Coiflets wavelet has high-order vanishing moments and good symmetry and is suitable for processing long-period smoothing characteristics of signals. It is designed to better preserve low-frequency information and is suitable for decomposing lower-frequency signal components. Long-term trends in PM2.5 data (such as seasonal fluctuations, monthly average concentration changes, etc.) are very suitable for using Coiflets wavelet. It can effectively decompose the long-period components and low-frequency trends of the signal, such as Coif1 and Coif3, which are often used for smooth signal analysis. Coiflets wavelet is suitable for analyzing seasonal and annual cycle fluctuations. It can be used to detect monthly trends and seasonal fluctuations in PM2.5 concentration (such as winter peaks and summer troughs). We use Coiflets wavelet to perform multi-level decomposition of PM2.5 time series data, and the results are shown in Fig. 3.7:

3.4.4.4 Biorthogonal Wavelet (biorN.N)

Biorthogonal wavelet (Cohen 1992) is a type of biorthogonal wavelet that can use different wavelet basis functions in signal decomposition and reconstruction. It has a linear phase characteristic and can reduce phase distortion in the process of signal decomposition and reconstruction. PM2.5 data analysis needs to retain the temporal structure of the signal and reduce the impact of phase distortion on the signal. Biorthogonal wavelets (such as bior4.4) perform well in signal compression and denoising, and can be used for smoothing and denoising of air quality data. Biorthogonal wavelets are suitable for denoising air quality monitoring data,

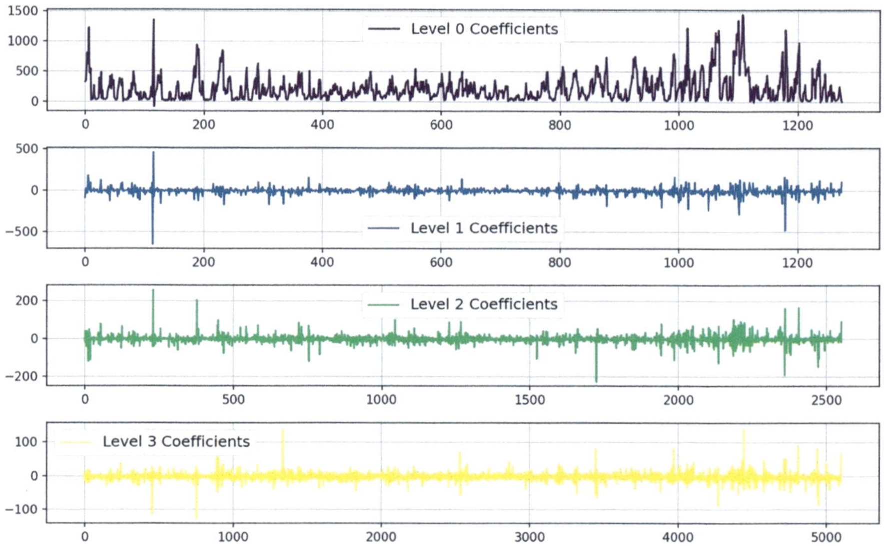

Fig. 3.8 bior4.4 wavelet decomposition—level 3

especially daily PM2.5 fluctuations and noise removal, and can be used for smoothing analysis and peak detection of PM2.5 data. We use Biorthogonal wavelet to perform multi-level decomposition of PM2.5 time series data, and the results are shown in Fig. 3.8:

We evaluate the performance of each level of wavelet classification and plot it in the Table 3.1:

The following conclusions can be drawn from the performance data of each wavelet function at different decomposition levels in the table: db4 and coif4 show very low MSE values at all levels, indicating that they are highly accurate in decomposing and reconstructing signals. In contrast, sym4 and bior4.4 have larger MSEs, indicating that they may be relatively weak in preserving signal details, especially at higher frequency decompositions.

The SNR values of db4 and coif4 are higher, especially when the SNR reaches more than 300 dB at level 3, indicating that these wavelets are effective in noise suppression. bior4.4 has the lowest SNR value, 242.12 dB at level 3, and further decreases in subsequent decomposition levels, showing its relatively low signal fidelity. The db4 and coif4 wavelets show high accuracy and noise suppression capabilities in signal decomposition and reconstruction and are suitable for scenarios that require high fine signal features. However, the sym4 and bior4.4 wavelets have relatively high MSE values and low SNR and may be more suitable for analysis tasks that tolerate large errors or require high frequency details. Depending on the specific application requirements, you can choose a suitable wavelet function to strike a balance between accuracy and noise suppression capabilities.

Table 3.1 The performance of each level of different wavelet classification

Wavelet function	Level	MSE	SNR	Correlation coefficient
db4	3	1.4087048038904179e-27	309.55 dB	1
	4	1.7922957082835778e-27	308.50 dB	1
	5	2.303921947175242e-27	307.41 dB	1
sym4	3	1.6245762625932698e-21	248.93 dB	1
	4	3.1228267268720512e-21	246.09 dB	1
	5	5.407932830189855e-21	243.70 dB	1
coif4	3	2.3953532075998377e-27	307.24 dB	1
	4	3.602910487585911e-27	305.47 dB	1
	5	4.401603963945719e-27	304.60 dB	1
bior4.4	3	7.785981718053387e-21	242.12 dB	1
	4	1.5764959239932202e-20	239.06 dB	1
	5	2.6857910365800675e-20	236.74 dB	1

3.4.5 Continuous Wavelet Transform

The Continuous Wavelet Transform (CWT) is a signal processing method used to analyze and characterize the non-stationary features of a signal through localized analysis in both time and frequency domains. CWT decomposes the signal into components at various scales and frequencies by scaling and translating the wavelet basis functions. This allows CWT to capture the frequency characteristics of the signal as they change over time, making it a powerful tool for time-frequency analysis. The core of CWT is to convolve the signal $x(t)$ with a wavelet function $\psi(t)$ to obtain the time-frequency representation of the signal. CWT is defined as follows:

$$C(a,b) = \int_{-\infty}^{+\infty} x(t) \cdot \psi_{a,b}^{*}(t)\,dt \tag{3.6}$$

where: $\psi_{a,b}(t)$ is the scaled and translated wavelet basis function, defined as:

$$\psi_{a,b}(t) = \frac{1}{\sqrt{|a|}}\psi\left(\frac{t-b}{a}\right) \tag{3.7}$$

The a is the scale parameter, which determines the width of the wavelet. Smaller values of a correspond to higher frequency details, while larger values correspond to lower frequency trends. b is the translation parameter that shifts the wavelet along the time axis. $*$ denotes the complex conjugate of the wavelet function.

By calculating $C(a, b)$ at different scales and positions, CWT can generate a time-frequency plane of the signal, thereby providing frequency information of the signal changing over time. In CWT, there is an inverse relationship between the scale parameter a and the frequency component. Smaller scales correspond to higher

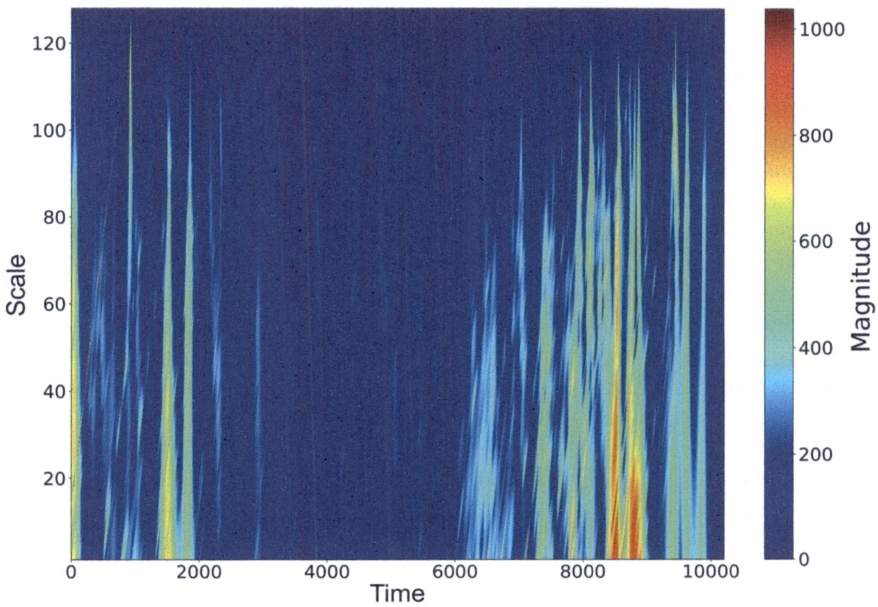

Fig. 3.9 Cwt scalogram of pm2.5 time series

frequencies, which can capture the high-frequency details of the signal, while larger scales correspond to low-frequency components, which are suitable for analyzing the long-term trend of the signal. This feature of CWT makes it very suitable for processing non-stationary signals, and it can provide different frequency resolutions in different periods. By displaying the wavelet coefficients $C\ (a,\ b)$ at different scales and time points on the time-frequency plane, CWT can generate a time-frequency diagram, where the brightness or color indicates the size of the coefficient. A large coefficient indicates that the signal has a stronger component at this time scale. The CWT power spectrum is the square of the wavelet coefficient, which is used to show the intensity changes of different frequency components over time. The results of the same time series data decomposition of the Temple of Heaven subway station are shown in Fig. 3.9:

The horizontal axis (Time) represents time, and the scale ranges from 0 to 10,000. This is the period for PM2.5 data sampling, and the characteristics of PM2.5 concentration changes over time can be seen. The vertical axis (Scale) represents the scale, ranging from 0 to 120. The larger the scale, the lower the corresponding frequency, indicating a longer-term trend of change; the smaller the scale, the higher the corresponding frequency, indicating a shorter-term change or high-frequency component. Color (Magnitude) The color represents the amplitude of the wavelet coefficient, and the amplitude comparison is given on the right side of the color bar. The color ranges from blue (low) to red (high), and the larger the value, the more obvious the fluctuation of PM2.5 concentration. The red area represents the part with higher signal energy at that time point and scale, while the blue area represents the part with lower energy.

There are multiple high-energy areas in red and yellow in the second half of the graph (around 6000 to 10,000 time points). This indicates that there were strong fluctuations in PM2.5 concentrations during these periods, possibly due to pollution events or other factors. Between scales 40 and 120, the energy is higher, which means that the changes during these periods are of lower frequency. For most of the period, the image is dominated by blue, indicating that the changes in PM2.5 concentrations are small. These areas may indicate a period of relatively stable PM2.5 concentrations. The fluctuations of PM2.5 vary at different scales. Some of the high-energy areas in the figure also appear at smaller scales (high frequency), indicating that PM2.5 concentrations have short-term rapid fluctuations in some periods. Large scales (such as 80–120) mainly capture overall trends and long-term changes.

For the low-frequency part, on a larger scale, higher amplitude values (such as red areas) indicate that PM2.5 concentration has a significant long-term change trend during these periods, which may indicate the influence of weather factors or seasonal changes.

For the high-frequency part, on a small scale, there are some areas of higher energy that show rapid changes in a short period. These may correspond to sudden pollution events, such as rapid increases in PM2.5 caused by industrial emissions, traffic congestion, etc.

3.5 Mode Decomposition of Air Quality Data

3.5.1 Empirical Mode Decomposition

Modal Decomposition (Huang et al. 1998) is a method used to decompose complex data into simpler modes. It is often applied to analyze different frequency and scale characteristics in a signal. Among the various modal decomposition methods, Empirical Mode Decomposition (EMD) is the most used. EMD is an adaptive data analysis method that can decompose time series signals into several intrinsic mode functions (IMFs) (Moore et al. 2018), remove noise components, and extract meaningful trends and periodic features. It is particularly suitable for processing nonlinear and non-stationary data, as it decomposes the original signal into multiple IMFs and a residual trend term to reveal the internal structure of the signal (Flandrin et al. 2004).

EMD is an adaptive method designed specifically for nonlinear and non-stationary signals. The core idea is to decompose the original signal into several intrinsic mode functions (IMFs) and a residual trend term. Each IMF satisfies the following two conditions:

- The difference between the number of extreme points and zero crossings does not exceed 1: For the entire data range, the difference between the number of extreme points (including maxima and minima) and the number of zero crossings in each IMF should not exceed 1. This ensures that the local frequency of the IMF remains relatively consistent within the signal range.
- The local average is zero: At any point in the IMF, the local average of its upper and lower envelopes (the envelopes obtained by spline interpolation of local maxima and minima) should be as close to zero as possible. This means that the IMF has symmetrical fluctuation characteristics and is suitable for describing periodic components in the signal.

The algorithm steps can be divided into the following five steps:

- Determine the extreme points: First, extract all local maxima and local minima from the original signal.
- Construct upper and lower envelopes: Perform cubic spline interpolation on the local maxima and minima to generate the upper and lower envelopes of the signal.
- Calculate the local average: Calculate the average of the upper and lower envelopes $m_1(t)$, that is, the local average of the signal.
- Extract detail components (candidate IMFs): Use $h1(t) = x(t) - m_1(t)$ to calculate the residual after subtracting the local mean m from the original signal x, and check whether h_1 satisfies the IMF condition. If it does, it is recorded as the first IMF as c_1. If not, continue to use h_1 as the new signal and repeat steps 1–4 until the IMF condition is met.
- Residual signal processing: Subtract the first IMF from the original signal to obtain the residual signal $r_1(t) = x(t) - c_1(t)$. Take r_1 as the new original signal and repeat the above steps to continue extracting the next layer of IMF until all IMFs are extracted.
- Termination condition: When the residual signal r_n becomes a monotonic function or contains a trend term, the EMD process terminates, and the final decomposition result is:

$$x(t) = \sum_{i=1}^{n} (c_i(t)) + r_n(t) \tag{3.8}$$

where $c_i(t)$ represents each IMF and $r_n(t)$ represents the residual trend term of the signal.

We performed EMD decomposition on the PM2.5 concentration time series data of the Temple of Heaven subway station within one year, and the intrinsic mode function obtained by the decomposition is shown in Figs. 3.10, 3.11, 3.12, and 3.13.

IMF 1: The highest frequency component, which fluctuates violently and changes rapidly, and has a random nature. IMF 1 may be caused by daily random noise or transient events (such as sudden emissions or sudden meteorological changes). The high-frequency components in this mode often reflect short-term, rapidly changing factors.

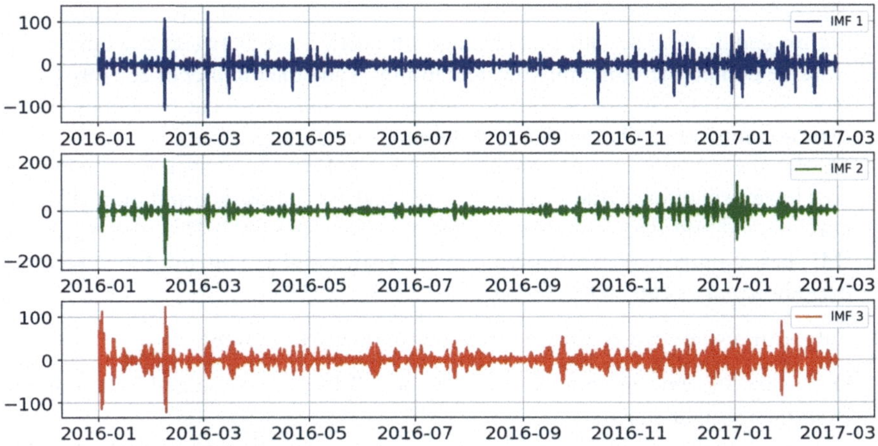

Fig. 3.10 IMF1, IMF2 and IMF3 of EMD Decomposition

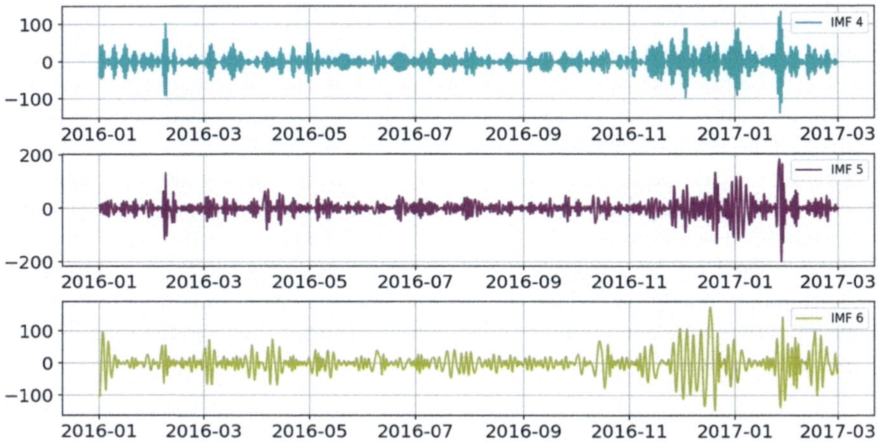

Fig. 3.11 IMF4, IMF5 and IMF6 of EMD Decomposition

IMF 2: The frequency is slightly lower than IMF 1, but it is still a high-frequency component with greater randomness. IMF 2 may also represent random changes in the short term or rapid changes in local environmental conditions, such as sudden weather changes or short-term industrial activities.

IMF 3: The frequency gradually decreases, showing a certain periodicity. IMF 3 may reflect periodic changes in weather conditions or short-term pollution sources, such as rainfall, wind speed changes, etc., which will affect PM2.5 concentrations in a short period.

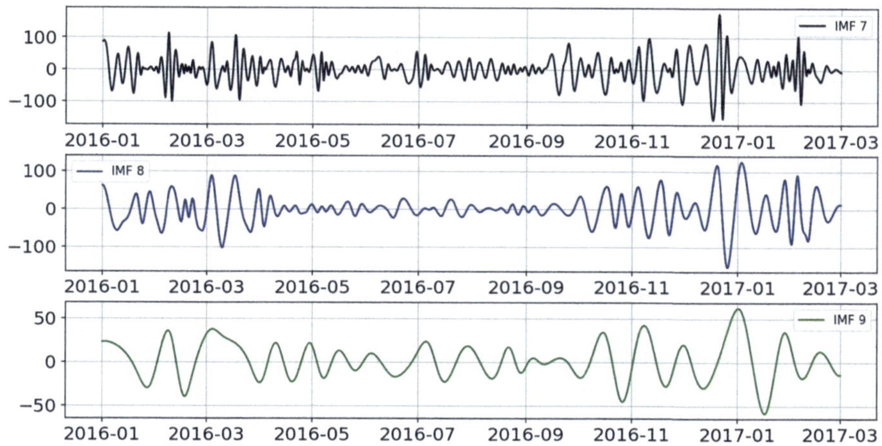

Fig. 3.12 IMF7, IMF8 and IMF9 of EMD Decomposition

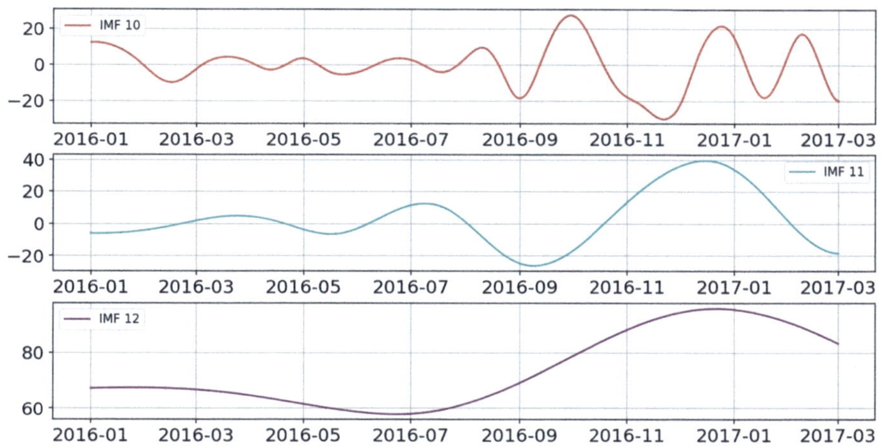

Fig. 3.13 IMF10, IMF11 and IMF12 of EMD Decomposition

IMF 4: A higher frequency component with violent fluctuations, showing greater randomness. IMF 4 may be caused by short-term weather events or local pollution sources. For example, the impact of short-term events such as strong winds and rainfall on PM2.5 concentrations.

IMF 5: Medium-to-high frequency, relatively violent fluctuations, but with certain regularity. This mode may correspond to environmental changes on a shorter time scale, such as daily traffic and short-term industrial emissions. These factors may cause rapid increases or decreases in PM2.5.

IMF 6: A lower frequency, with a cycle of about 2 weeks and relatively stable fluctuations. IMF 6 may be related to weekly changes, such as the difference in emissions on weekends and weekdays. This fluctuation may reflect the improvement effect of reduced industrial production and traffic activities on weekends on air quality.

IMF 7: Medium frequency, small amplitude, cycle of about 1 month, and relatively smooth fluctuations. IMF 7 may correspond to monthly trends, such as weather patterns in a particular month or monthly activities (such as agricultural burning, factory emissions).

IMF 8: Slightly lower frequency than IMF 7, with a cycle of between 1 and 2 months and larger amplitude. IMF 8 may reflect climate or atmospheric transmission factors that affect PM2.5 concentrations. For example, monthly changes in wind speed and humidity may affect the dispersion of pollutants.

IMF 9: Lower frequency, cycle of about 3 months, and smooth fluctuations. IMF 9 is related to seasonal changes and may reflect seasonal changes in atmospheric conditions, such as seasonal effects of monsoons or temperatures.

IMF 10: Low frequency, medium amplitude, showing relatively regular fluctuations. Its period is about 1–2 months. This mode may correspond to seasonal factors, reflecting medium-term fluctuations in PM2.5 concentrations.

IMF 11: Lower frequency, with a period close to 4–6 months and larger amplitude. IMF 11 may reflect seasonal changes on a longer time scale. For example, heating in winter increases PM2.5 concentrations, while it decreases relatively in summer.

IMF 12: The lowest frequency mode, with a period of more than one year, shows obvious trend changes. IMF 12 represents a long-term trend and may reflect the overall trend of air quality changes within a year, which is related to changes in overall policies or economic activities.

3.5.2 Variations and Improvements of the Traditional EMD Method

Ensemble empirical mode decomposition (EEMD) is an improved method developed on the basis of the traditional EMD method, which is used to solve the common "mode mixing" problem in EMD decomposition. Specifically, the EEMD method adds white noise to the signal multiple times, performs EMD decomposition on each signal after adding noise, and finally averages all the decomposition results to obtain a more stable intrinsic mode function (IMF). Mode aliasing refers to the presence of multiple different frequency components in an intrinsic mode function (IMF), or similar frequency components appearing in different IMFs (Wu and Huang 2009), which makes the decomposition results difficult to interpret and use. EEMD introduces white noise into the signal and uses the statistical characteristics of white noise to separate different frequency components, thereby reducing the mode aliasing phenomenon.

Specifically, EEMD adds white noise of different amplitudes to the signal and performs EMD decomposition on the signal with each noise addition. Since each added noise has different characteristics, the noise component will interfere with the mode in the decomposition, so that the aliased frequency components in the intrinsic mode function tend to average after multiple decompositions. All decomposition results are averaged to obtain the final result of each IMF, making the mode clearer and more stable. Complete Ensemble Empirical Mode Decomposition (CEEMD) is an improvement on EEMD. It adds noise symmetry and offsets the residual effect of noise by adding opposite noise pairs (positive and negative noise), the way of adding noise in EEMD is improved to make the decomposition of each mode more stable, thereby further improving the decomposition accuracy (Torres et al. 2011).

The IMF spectra after EMD, EEMD and CEEMD decomposition are shown in Figs. 3.14, 3.15, and 3.16.

3.5.2.1 EMD Spectrum Separation

From the spectrum distribution, we can see that the IMF decomposed by EMD has a higher amplitude in the low frequency band and some fluctuations in the higher frequency band. Since EMD does not perform direct optimization in the frequency domain, mode aliasing is prone to occur. The spectra of different IMFs overlap a lot and are difficult to distinguish. In the IMF decomposed by EMD, frequency

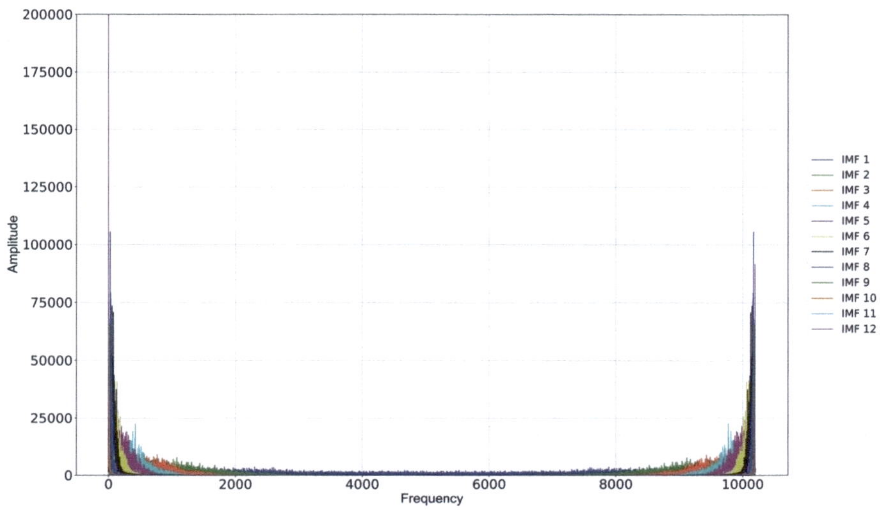

Fig. 3.14 Spectrum Separation of EMD IMFs

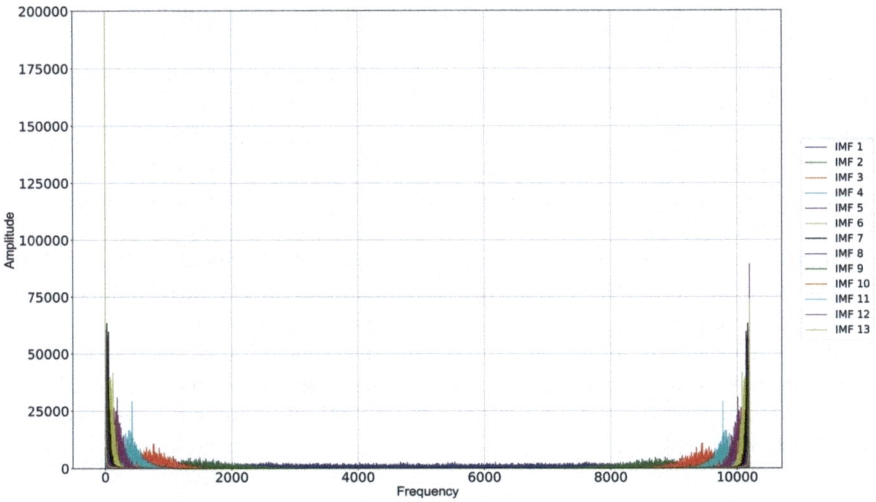

Fig. 3.15 Spectrum Separation of EEMD IMFs

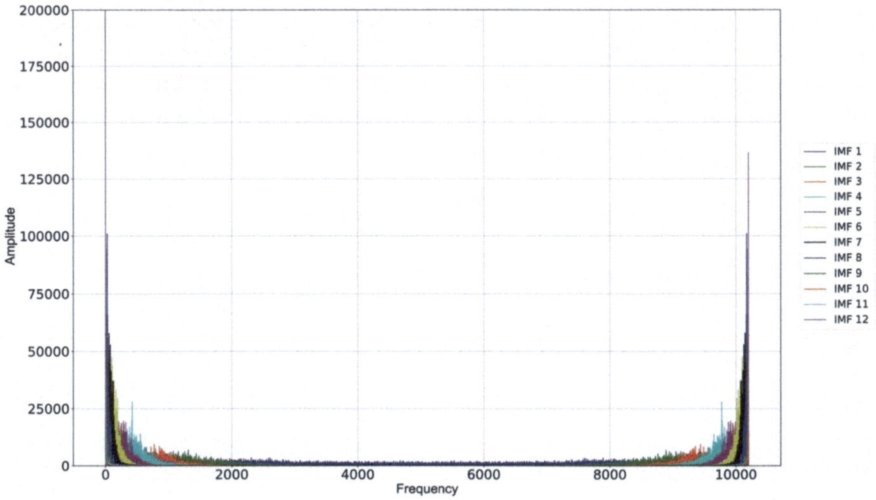

Fig. 3.16 Spectrum Separation of CEEMD IMFs

components are easily mixed together, that is, similar frequencies may be contained in multiple IMFs, which means that there is mode aliasing in the time domain. This situation is more obvious in complex signals. Although EMD can extract the basic mode of the signal, due to the lack of noise perturbation and smoothing averaging processing, the frequency characteristics of each IMF are difficult to completely separate and clearly, especially when spectrum overlap occurs in the low-frequency area.

3.5.2.2 EEMD Spectrum Separation

In the EEMD decomposition diagram, the spectrum separation of IMF is improved compared to EMD. Since EEMD adds white noise and performs multiple decomposition and averaging during the decomposition process, it partially suppresses mode aliasing, making each IMF more independent. EEMD makes the spectrum distribution clearer and the separation between frequency components is enhanced through the disturbance effect of white noise. The spectrum crossover between low-frequency and high-frequency IMFs is reduced, and the spectrum energy is more concentrated. EEMD better suppresses mode aliasing, provides better modal separation when decomposing complex signals, and makes the physical meaning of each IMF clearer.

3.5.2.3 CEEMD Spectrum Separation

In the CEEMD diagram, it can be observed that the spectrum distribution is clearer than that of EMD and EEMD. CEEMD uses positive and negative noise pairs to cancel each other out, thereby further reducing the influence of residual noise on the basis of multiple decomposition and averaging. CEEMD has the best spectrum separation effect. By eliminating the influence of residual noise, the IMFs of CEEMD are more concentrated on the spectrum, and the frequency characteristics of each IMF are clearly separated. There is less frequency crossover, and the frequency boundaries between low-frequency and high-frequency IMFs are clear. CEEMD further improves the accuracy of spectrum separation and has a higher decomposition stability. The frequency components of each IMF are relatively independent, which helps to more accurately analyze different modes in the signal.

3.6 Decomposition Performance Comparison

When comparing the decomposition performance of EMD, EEMD, CEEMD, linear wavelet transforms, discrete wavelet transforms and their variants, a comprehensive comparison can be made from the following key parameters and performance indicators:

3.6.1 Decomposition Accuracy

For each decomposition method, the reconstructed signal is compared with the original signal. Through multiple decomposition experiments, the average reconstruction error and SNR of each decomposition can be calculated to generate statistical results for comparative analysis of the decomposition accuracy of each method. The performance of different data decomposition methods is shown in the following Table 3.2.

Table 3.2 Performance of different data decomposition methods

Decomposition method	Level	MSE	SNR
db4	3	1.4087048038904179e-27	309.55 dB
	4	1.7922957082835778e-27	308.50 dB
	5	2.303921947175242e-27	307.41 dB
sym4	3	1.6245762625932698e-21	248.93 dB
	4	3.1228267268720512e-21	246.09 dB
	5	5.407932830189855e-21	243.70 dB
coif4	3	2.3953532075998377e-27	307.24 dB
	4	3.602910487585911e-27	305.47 dB
	5	4.401603963945719e-27	304.60 dB
bior4.4	3	7.785981718053387e-21	242.12 dB
	4	1.5764959239932202e-20	239.06 dB
	5	2.6857910365800675e-20	236.74 dB
EMD		5.663711942269131e-30	SNR: 333.50 dB
EEMD		7419.133859989786	2.33 dB
CEEMD		2.1682104933634585e-28	SNR: 317.67 dB

db4 and coif4 wavelets have lower MSE and higher SNR, and are suitable for high-precision data decomposition tasks. Sym4 and bior4.4 wavelets have relatively high MSE, low SNR, and poor decomposition accuracy. And as the decomposition level increases, the decomposition error will increase. In comparison, EMD and CEEMD are significantly better than the wavelet method in terms of accuracy. EMD shows the lowest MSE and the highest SNR, and is suitable for extremely high-precision decomposition tasks. The lower accuracy of EEMD may be due to the fact that in EEMD, the averaging operation of each decomposition result cannot completely offset the influence of noise, especially when the amount of noise is large or the signal itself is complex, which may lead to a higher MSE thus affecting the accuracy.

EEMD is effective in alleviating modal aliasing, but the introduction of superposition and accumulation effects of noise will reduce its decomposition accuracy, which is manifested in higher MSE and lower SNR. The decomposition performance of EEMD is highly dependent on the noise amplitude and the number of iterations, and improper parameter selection will significantly affect the decomposition accuracy. In situations where high accuracy is required, EMD and CEEMD may be more suitable because they retain the fine features of the signal better and are more suitable for decomposition tasks with high accuracy requirements.

3.6.2 *Computational Complexity*

The computational complexity of modal decomposition is relatively high, especially for EEMD and CEEMD, which require multiple decompositions to eliminate residual noise, so the computational time is relatively long. Wavelet decomposition is computationally efficient, suitable for real-time processing, and has good computational performance for long signals. Modal decomposition requires a large amount of computation when processing long time series or large-scale data, while wavelet decomposition can be decomposed more efficiently due to its fast algorithm.

3.6.3 Boundary Effect

Modal decomposition (such as EMD, EEMD, and CEEMD) relies on the local extreme points of the signal for decomposition. However, at the boundary (starting point and end point) of the signal, the number of extreme points is small or cannot extend beyond the boundary, resulting in the inability to accurately extract the local features of the decomposition calculation. Due to the lack of sufficient data support at the boundary of the signal, the calculation of the envelope and IMF becomes unstable, resulting in inaccurate decomposition results at the boundary position, or even distortion. As a basic method of modal decomposition, EMD is most obvious in terms of boundary effects. The extreme points at the boundary are not enough to support a smooth envelope, resulting in large fluctuations in the IMF at the boundary. EEMD partially alleviates the boundary effect by adding white noise to disturb the signal, performing multiple decompositions and averaging. The interference of white noise makes the fluctuation distribution at the boundary more uniform to a certain extent, so the decomposition result at the boundary is slightly improved, but the boundary effect still exists. CEEMD adds positive and negative noise pairs on the basis of EEMD. The average result obtained through multiple decompositions is smoother and the boundary effect is further reduced. The positive and negative noise pairs can eliminate some noise interference on the boundary, making the decomposition result more stable.

For wavelet decomposition, since the convolution operation requires the signal to have a certain length, the wavelet cannot completely cover the entire window at the boundary of the signal. During the convolution process, the data at the boundary will be lost, resulting in inaccurate decomposition results at the boundary position. This inaccuracy is reflected as boundary distortion, that is, abnormal fluctuations or discontinuities appear at the start and end positions of the signal. Wavelet transform can effectively alleviate the distortion caused by boundary effect through signal filling and boundary extension strategy. Compared with modal decomposition, the boundary processing of wavelet transform is more flexible and stable.

The performance of data decomposition methods can be compared from three aspects: decomposition accuracy, computational complexity and boundary effects. In terms of decomposition accuracy, EMD and CEEMD show higher accuracy with lower reconstruction error and higher signal-to-noise ratio and are therefore suitable for high-precision data decomposition tasks. In contrast, EEMD is prone to higher errors in complex signals due to insufficient noise cancellation caused by noise superposition. In wavelet decomposition, db4 and coif4 wavelets have better accuracy, while sym4 and bior4.4 perform worse. In addition, the increase in the number of decomposition layers will lead to the accumulation of errors and reduce the decomposition accuracy. In terms of computational complexity, modal decomposition (such as EEMD and CEEMD) requires multiple repeated decompositions, so the calculation amount is large and the time is long, while the wavelet decomposition algorithm is more efficient and suitable for real-time processing and decomposition of long signals.

Secondly, in terms of boundary effects, modal decomposition relies on local extreme points of the signal, so distortion and inaccuracy are prone to occur at the boundaries. The boundary effect of EMD is the most significant, and EEMD partially improves this problem by adding white noise, but the boundary effect still exists. CEEMD further reduces the instability of the boundary through positive and negative noise pairs, making the decomposition results smoother and more stable. In contrast, wavelet decomposition uses signal filling and boundary extension strategies in boundary processing, which makes wavelet more flexible and stable in dealing with boundary effects. Therefore, wavelet decomposition has certain advantages over modal decomposition in processing signal boundary stability.

3.7 Conclusions

This chapter mainly discusses the application of data decomposition technology in air quality monitoring, focusing on wavelet decomposition and modal decomposition methods. These technologies can effectively handle the nonlinear, non-stationary, and multi-scale characteristics of air quality data. Through multi-scale decomposition, long-term trends, seasonal changes, and random fluctuations can be extracted to provide data support for policymaking and environmental governance. Wavelet decomposition is suitable for multi-resolution analysis and noise reduction, and wavelet bases such as Daubechies are commonly used; modal decomposition (such as EMD, EEMD, CEEMD) can adaptively extract intrinsic mode functions (IMFs) and reveal the multi-frequency characteristics of complex signals. The comparison shows that modal decomposition is superior to wavelet decomposition in terms of accuracy and signal restoration, but the computational complexity is high, and the boundary effect is obvious. The study also shows the advantages and disadvantages of different methods and applicable scenarios and verifies the decomposition effect in combination with PM2.5 data analysis cases, providing a comprehensive technical reference for air quality monitoring.

References

Anenberg SC, West JJ, Yu HB, Chin M, Schulz M, Bergmann D, Bey I, Bian HS, Diehl T, Fiore A, Hess P, Marmer E, Montanaro V, Park R, Shindell D, Takemura T, Dentener F (2014) Impacts of intercontinental transport of anthropogenic fine particulate matter on human mortality. Air Qual Atmos Health 7(3):369–379. https://doi.org/10.1007/s11869-014-0248-9

Brown Ingram J (2009) A wavelet tour of signal processing: the sparse way. Invest Oper 30(1):85–87

Chavan MS, Mastorakis N, Chavan MN, Gaikwad MS (2011) Implementation of SYMLET wavelets to removal of Gaussian additive noise from speech signal. In: Proceedings of recent researches in communications, automation, signal processing, nanotechnology, astronomy and nuclear physics: 10th WSEAS international conference on electronics, hardware, wireless and optical communications (EHAC'11), Cambridge, p 37

Cohen A (1992) Biorthogonal wavelets. In: Wavelets: a tutorial in theory and applications, vol 2, pp 123–152

Cohen R (2012) Signal denoising using wavelets. Project report, Department of Electrical Engineering Technion, Israel Institute of Technology, Haifa, pp 890

Daubechies I (1988) Orthonormal bases of compactly supported wavelets. Commun Pure Appl Math 41(7):909–996

Dautov CP, Özerdem MS (2018) Introduction to wavelets and their applications in signal denoising. Bitlis Eren Univ J Sci Technol 8(1):1–10

Flandrin P, Rilling G, Goncalves P (2004) Empirical mode decomposition as a filter bank. IEEE Signal Process Lett 11(2):112–114

Fujiwara T, Shilpika S, Sakamoto N, Nonaka J, Yamamoto K, Ma KL (2021) A visual analytics framework for reviewing multivariate time-series data with dimensionality reduction. IEEE Trans Vis Comput Graph 27(2):1601–1611. https://doi.org/10.1109/tvcg.2020.3028889

Guo Q, He Z, Wang Z (2023a) Predicting of daily PM2.5 concentration employing wavelet artificial neural networks based on meteorological elements in Shanghai, China. Toxics 11(1):51. https://doi.org/10.3390/toxics11010051

Guo Q, He Z, Wang Z (2023b) Simulating daily PM2.5 concentrations using wavelet analysis and artificial neural network with remote sensing and surface observation data. Chemosphere 340:139886. https://doi.org/10.1016/j.chemosphere.2023.139886

Huang NE, Shen Z, Long SR, Wu MC, Shih HH, Zheng Q, Yen NC, Tung CC, Liu HH (1998) The empirical mode decomposition and the Hilbert spectrum for nonlinear and non-stationary time series analysis. Proc R Soc Lond A 454(1971):903–995

Klionskiy D, Kupriyanov M, Kaplun D (2017) Signal denoising based on empirical mode decomposition. J Vibroeng 19(7):5560–5570

Lang WC, Forinash K (1998) Time-frequency analysis with the continuous wavelet transform. Am J Phys 66(9):794–797

Luo X, Jiang R, Yang B, Qin H, Hu H (2024) Air quality visualization analysis based on multivariate time series data feature extraction. J Vis 27(4):567–584. https://doi.org/10.1007/s12650-024-00981-3

Moore KJ, Kurt M, Eriten M, McFarland DM, Bergman LA, Vakakis AF (2018) Wavelet-bounded empirical mode decomposition for measured time series analysis. Mech Syst Signal Process 99:14–29

Niu C, Niu Z, Qu Z, Wei L, Li Y (2022) Research and application of the mode decomposition-recombination technique based on sample-fuzzy entropy and K-means for air pollution forecasting. Front Environ Sci 10:941405. https://doi.org/10.3389/fenvs.2022.941405

Rhif M, Ben Abbes A, Farah IR, Martínez B, Sang Y (2019) Wavelet transform application for/in non-stationary time-series analysis: a review. Appl Sci 9(7):1345

Sharif I, Khare S (2014) Comparative analysis of Haar and Daubechies wavelet for hyper spectral image classification. Int Arch Photogramm Remote Sens Spat Inf Sci XL-8:937–941. https://doi.org/10.5194/isprsarchives-XL-8-937-2014

Sifuzzaman M, Islam MR, Ali MZ (2009) Application of wavelet transform and its advantages compared to Fourier transform

Silik A, Noori M, Altabey WA, Ghiasi R, Wu Z (2021) Comparative analysis of wavelet transform for time-frequency analysis and transient localization in structural health monitoring. Struct Durab Health Monit 15(1):1

Torres ME, Colominas MA, Schlotthauer G, Flandrin P (2011) A complete ensemble empirical mode decomposition with adaptive noise. In: 2011 IEEE international conference on acoustics, speech and signal processing (ICASSP). IEEE, pp 4144–4147

Vetterli M (1995) Wavelets and subband coding. Prentice Hall
Wei D, Bovik AC, Evans BL (1997) Generalized coiflets: a new family of orthonormal wavelets. In: Conference record of the thirty-first Asilomar conference on signals, systems and computers (Cat. No. 97CB36136). IEEE, pp 1259–1263
Wu Z, Huang NE (2009) Ensemble empirical mode decomposition: a noise-assisted data analysis method. Adv Adapt Data Anal 1(1):1–41
Yang BY, Liu YM, Hu LW, Zeng XW, Dong GH (2017) Urgency to assess the health impact of ambient air pollution in China. In: Dong GH (ed) Ambient air pollution and health impact in China, vol 1017. Advances in experimental medicine and biology, pp 1–6. https://doi.org/10.1007/978-981-10-5657-4_1

Chapter 4
Data Identification in Air Quality Monitoring

Abstract This chapter delves into the significance of data identification in air quality monitoring, focusing on techniques that can enhance the accuracy of data analysis and forecasting. With the rise of complex air quality datasets encompassing various pollutants and meteorological factors, effective data identification has become essential to filter out noise and extract meaningful patterns. The chapter reviews two primary methods: feature selection and feature extraction. Feature selection emphasizes identifying the most impactful variables, while feature extraction transforms raw data to capture key trends better. A comparative analysis of feature selection methods—filter and wrapper—demonstrates the superior predictive accuracy of the wrapper method, particularly in multistep forecasting. Additionally, performance evaluations of statistical feature extraction and time-frequency analysis (via DWT) reveal the unique advantages of each approach for different prediction scenarios, thereby underscoring the importance of tailored data identification strategies for optimal air quality forecasting.

4.1 Introduction

Air quality monitoring is a critical component in ensuring environmental sustainability and public health. In recent years, technological advancements have significantly enhanced our ability to collect vast amounts of air quality data through various monitoring stations, satellites, and sensors (Singh et al. 2021). However, the effective utilization of this data relies heavily on our capacity to accurately identify relevant patterns, anomalies, and relationships within the datasets (Thudumu et al. 2020). This process, known as data identification, is a vital step in transforming raw data into actionable insights, enabling improved decision-making in air quality management (Morabito and Versaci 2003).

H. Liu et al., *Data Science in Air Quality Monitoring*, Engineering Applications of Computational Methods 23, https://doi.org/10.1007/978-981-96-5777-3_4

Data identification refers to the process of selecting and isolating relevant information from large and complex datasets, particularly in cases where multiple variables and factors influence the outcome (Belsley et al. 2005). In the context of air quality monitoring, data identification involves recognizing key pollutants, identifying patterns in pollution levels, and correlating these patterns with external factors such as weather conditions, traffic, and industrial activities (Austin et al. 2012). This step is crucial as it helps to separate noise from meaningful information, allowing for more accurate predictions and analysis of air quality trends (Rabie et al. 2024).

This chapter will explore the techniques and methods used for data identification in air quality monitoring. These methods include feature selection, which is the process of identifying the most important variables in a dataset, and feature extraction, which involves transforming the raw data into a format that is easier to analyze. Furthermore, we will examine the performance comparison of different identification methods to determine their effectiveness in various air quality monitoring scenarios.

4.1.1 The Importance of Data Identification in Air Quality Monitoring

Air quality data is typically vast and multidimensional, containing information from a variety of sensors and sources. For instance, typical air quality datasets may include measurements of pollutants such as PM2.5, PM10, ozone (O_3), nitrogen dioxide (NO_2), sulfur dioxide (SO_2), carbon monoxide (CO), and volatile organic compounds (VOCs) (Breuer 1999). In addition to these pollutants, data from meteorological stations (e.g., temperature, humidity, wind speed) and socioeconomic variables (e.g., traffic data, industrial output) are often integrated to provide a more comprehensive understanding of the factors affecting air quality (Tian et al. 2020).

However, this diversity of data also presents significant challenges. Not all variables in these datasets are equally important, and some may even introduce noise, reducing the effectiveness of predictive models and analyses (Hasan and Chu 2022). The task of identifying which variables are most relevant—through feature selection—becomes essential for improving the quality of air quality forecasting and intervention strategies. Additionally, by reducing the dimensionality of the data through feature extraction, we can focus on the most critical variables, making it easier to detect pollution patterns and predict air quality outcomes (Li et al. 2017).

Data identification is also important for ensuring that air quality models are interpretable and understandable for policymakers and stakeholders. Without effective data identification, the sheer volume and complexity of air quality data can make it difficult to draw clear conclusions or make informed decisions.

4.1.2 *Methods for Data Identification in Air Quality Monitoring*

Air quality data are primarily time-series data, collected over continuous intervals and influenced by a multitude of dynamic factors such as weather conditions, industrial activities, and traffic patterns (Yang and Wang 2017; Carslaw and Ropkins 2012; Baldasano et al. 2003). Effectively identifying meaningful patterns in such time-series datasets is essential for predicting pollution levels, detecting anomalies, and making informed decisions for environmental management. Data identification in air quality monitoring entails a systematic approach to identifying patterns, relationships, and trends in the data. The methods of data identification are mainly divided into feature selection and feature extraction (Zebari et al. 2020).

4.1.2.1 Feature Selection

Feature selection is the process of selecting the most relevant variables in a dataset while discarding redundant or irrelevant information (Zebari et al. 2020). For air quality monitoring, this might involve selecting key pollutants that have the most significant impact on air quality indices or human health outcomes. There are various approaches to feature selection, including (Jović et al. 2015):

- Filter methods: These methods rank each feature by its statistical relationship with the output variable, such as using correlation coefficients or mutual information (Yang and Wang 2017). For example, selecting variables that show the highest correlation with pollution levels can help reduce the dimensionality of the dataset while maintaining important information.
- Wrapper methods: These methods evaluate different subsets of features and select the combination that produces the best performance for a specific predictive model (El Aboudi and Benhlima 2016). This approach is often computationally intensive but can provide more accurate results than filter methods. In the context of air quality, this might involve selecting features that yield the highest accuracy for forecasting pollution levels.
- Embedded methods: These methods incorporate feature selection as part of the model training process (Lal et al. 2006). For instance, decision tree-based algorithms such as Random Forest or eXtreme Gradient Boosting (XGBoost) naturally perform feature selection during model training by evaluating which features contribute the most to predicting the outcome.

4.1.2.2 Feature Extraction

While feature selection focuses on selecting the most relevant existing variables, feature extraction involves transforming raw data into new features that are more suitable for analysis (Zebari et al. 2020). In air quality monitoring, feature extraction can be used to create new variables that better capture patterns in pollution data. Some of the most used techniques for feature extraction include:

- Statistical feature extraction: This method involves calculating statistical metrics from the air quality data, such as mean, variance, maximum, minimum, median, and standard deviation (Mutlag et al. 2020). These features describe the distribution and variation trends in the time series data, providing a simple and direct approach suitable for preliminary analysis and as inputs for models.
- Time-frequency analysis: Time-frequency methods analyze both the time-domain and frequency-domain characteristics of the data (Mutlag et al. 2020). For example, Fourier transform converts the time series into the frequency domain, helping to analyze periodicity and frequency components. Wavelet transform is also widely used, capable of extracting both time-domain and frequency-domain features simultaneously, making it suitable for capturing short-term fluctuations and long-term trends in air quality data.

4.2 Data Acquisition

The data used in this chapter is identical to that in Chap. 2, originating from 13 major cities in the Jing-Jin-Ji region, including Beijing, Tianjin, Shijiazhuang, Zhangjiakou, Chengde, Qinhuangdao, Tangshan, Baoding, Langfang, Cangzhou, Hengshui, Xingtai, and Handan. As a heavy industrial hub, the air quality issues in Jing-Jin-Ji have garnered significant attention, especially due to the concentration of steel and coal industries in the area. The emission of pollutants presents complex spatial and temporal characteristics. To conduct in-depth air quality monitoring analysis, this chapter focuses on the time-series data of seven key air pollutants (AQI, PM2.5, PM10, SO_2, NO_2, O_3, and CO) and performs feature identification and modeling.

Unlike Chap. 2, where the emphasis was on data preprocessing and basic time-series descriptions, this chapter delves deeper into uncovering underlying patterns in the data to support the development of models for pollutant identification and prediction. The dataset includes 36,000 samples spanning from May 2014 to June 2018, with a 1-h interval. Compared to static data, time-series data exhibits continuity and dynamic variations, providing more potential information for feature extraction and pattern recognition. The primary objective of this chapter is to utilize these time-series data, applying scientific methods to achieve precise pollutant feature identification, thereby improving the reliability and effectiveness of subsequent prediction models.

One major challenge in the study lies in the strong non-stationarity and periodicity present in the pollutant time-series data. For example, pollutant concentrations often fluctuate periodically due to seasonal meteorological conditions, industrial activities, and regulatory interventions (Afifa et al. 2024). These characteristics necessitate careful consideration of time-varying features and potential patterns across different time intervals during the processes of feature selection and extraction. Additionally, the high correlations between different pollutants introduce complexity in extracting and analyzing multidimensional features (Luo et al. 2024). Although averaging data across monitoring stations for each city helps reduce fluctuations from individual stations, effective algorithms are still required to uncover inter-pollutant relationships and enhance identification accuracy.

4.3 Feature Selection of Air Quality Data

4.3.1 Feature Selection Performance Evaluation

To evaluate the effectiveness of feature selection methods, data from four cities—Beijing, Tianjin, Shijiazhuang, and Zhangjiakou—were selected, labeled as series #A1, A2, A3, and A4, respectively. Each time series consists of 1500 samples, including seven features: AQI, PM2.5, PM10, SO_2, NO_2, O_3, and CO. For performance comparison, the first 1000 samples are used as the training set, and the remaining 500 samples as the test set.

In the process of feature selection, it is crucial to define the target variable to be predicted, as the selection is based on the correlation or contribution between features and the target. In this chapter, the target variable for feature selection is AQI, and the AQI time series for the four series is shown in Fig. 4.1. After applying feature selection methods, AQI prediction is performed using a classic model—Long Short-Term Memory network (LSTM)—based on the selected features (Yu et al. 2019). The prediction results will be compared using evaluation metrics such as MSE, RMSE, and MAE to assess the performance of different feature selection methods. The prediction time window is set to 24 h, and the prediction step is defined as n, meaning that data from hour $x-23$ to hour x are used to predict AQI at hour $x + n$, denoted as n-step. The time window schematic is shown in Fig. 4.2.

In this chapter's performance evaluation, Root Mean Square Error (RMSE), Mean Absolute Error (MAE), Pearson correlation coefficient (P), and Kling–Gupta Efficiency (KGE) were chosen as evaluation metrics. These four indicators comprehensively measure the accuracy and stability of the prediction model from different perspectives. RMSE and MAE are common standards for measuring error; RMSE is more sensitive to larger errors, while MAE assigns equal weight to all errors. P, as a standardized correlation metric, quantitatively evaluates the linear correlation between predicted and actual values, with a value closer to 1 indicating a stronger correlation. KGE is a comprehensive indicator that not only considers the correlation between predicted and actual values but also assesses the bias and variability ratio of the predictions, thus providing a more holistic evaluation of model performance. The combined use of these metrics ensures a multidimensional evaluation of the feature selection methods.

The equations of the MAE, RMSE, P and KGE are presented as follows:

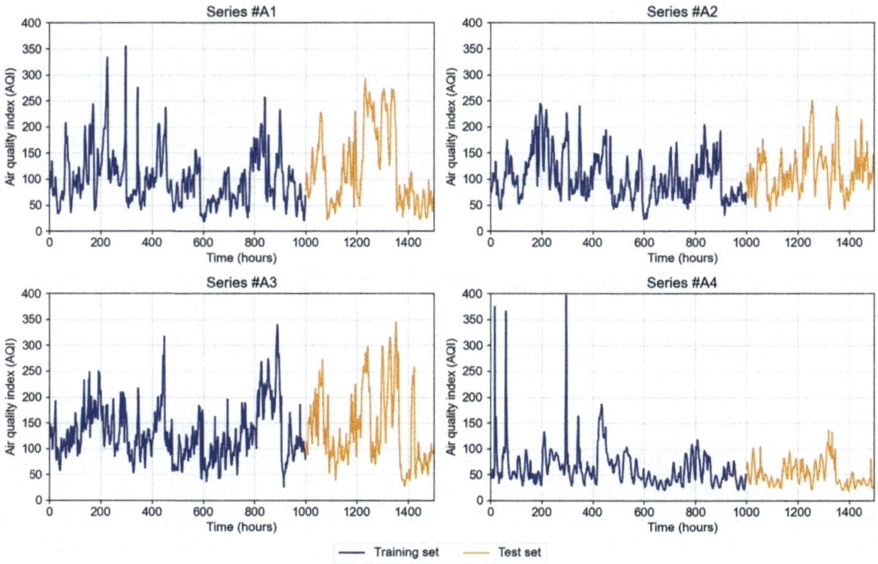

Fig. 4.1 AQI comparison among four cities

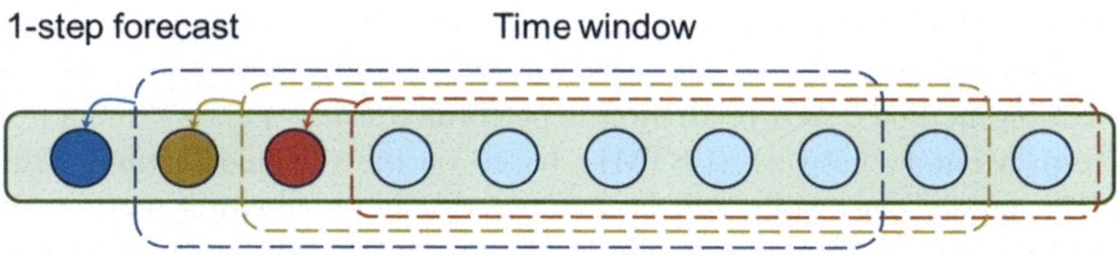

Fig. 4.2 The diagram of time window

$$\mathrm{MAE} = \left(\sum_{t=1}^{N} \left| Y_t - \hat{Y}_t \right| \right) / N \tag{4.1}$$

$$\mathrm{RMSE} = \sqrt{\frac{1}{N} \sum_{t=1}^{N} \left(Y_t - \hat{Y}_t \right)^2} \tag{4.2}$$

$$P = \frac{\mathrm{cov}\left(Y, \hat{Y}\right)}{\sigma_Y \sigma_{\hat{Y}}} \tag{4.3}$$

$$\mathrm{KGE} = 1 - \sqrt{\left(P-1\right)^2 + \left(\frac{\sigma_{\hat{Y}}}{\sigma_Y} - 1\right)^2 + \left(\frac{\mu_{\hat{Y}}}{\mu_Y} - 1\right)^2} \tag{4.4}$$

where Y and $\hat{Y}$ are the actual and estimated values respectively; $\mathrm{cov}\left(Y, \hat{Y}\right)$ is the covariance between the Y and $\hat{Y}$; σ_Y and $\sigma_{\hat{Y}}$ are the standard deviations of the Y and $\hat{Y}$; μ_Y and $\mu_{\hat{Y}}$ are the standard deviations of the Y and $\hat{Y}$.

4.3.2 *Filter Methods*

4.3.2.1 Theoretical Basis

In filter methods, one of the most commonly used techniques is the Pearson correlation coefficient (Gong et al. 2024). The calculation formula for the Pearson correlation coefficient is shown in the eq. (4.3). This method ranks features by measuring the linear correlation between each feature and the target variable. The Pearson correlation coefficient, denoted as P, takes values between -1 and 1, where values close to 1 or -1 indicate a strong positive or negative correlation, respectively, and values close to 0 indicate little or no correlation. Researchers typically select features with high correlation to the target variable and remove those with low or near-zero correlation. This helps in reducing the dimensionality of the dataset while retaining the most valuable information for predicting pollution levels.

The currently accepted guidelines for interpreting the correlation coefficient are shown in Table 4.1 (Ratner 2009). The Pearson correlation measures the strength and direction of a linear relationship between two variables. If $P = 1$, it indicates a perfect positive linear relationship, meaning as one variable increases, the other also increases proportionally. If $P = -1$, there is a perfect negative linear relationship, meaning as one variable increases, the other decreases. If $P = 0$, it indicates no linear relationship between the variables. In feature selection, this coefficient helps in identifying which features are most relevant to the target variable, as strong correlations imply that the feature has more predictive power. This method is particularly useful in large datasets where selecting the most relevant features can improve model performance and computational efficiency.

4.3.2.2 Modeling Step

In the analysis of the four series, the correlation coefficients between the AQI and other features are illustrated in Fig. 4.3. Given the variability in correlation coefficient thresholds, it is essential to investigate the impact of different feature selection criteria on the predictive performance. As shown in Table 4.1, we selectively choose features with correlation coefficients greater than or equal to 0.0, 0.3, and 0.7 for the purpose of time series forecasting, as well as AQI itself. This structured approach

Table 4.1 General guidelines for the interpretation of Pearson correlation coefficient

Coefficient	Interpretation
$\lvert P\rvert = 0.0$	No relationship
$0.0 < \lvert P\rvert < 0.3$	Weak relationship
$0.3 \leq P\rvert < 0.7$	Moderate relationship
$0.7 \leq \lvert P\rvert < 1.0$	Strong relationship
$\lvert P\rvert = 1.0$	Perfect relationship

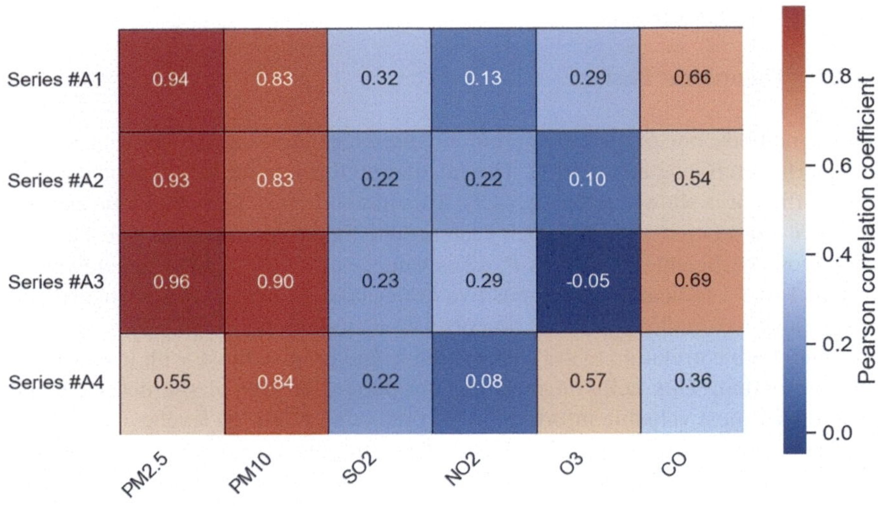

Fig. 4.3 Correlation coefficient between AQI and other features

enables us to assess how varying thresholds influence the effectiveness of the selected features in predicting AQI values.

In the 1-step prediction using LSTM, the scatter plots of predicted AQI versus actual AQI are shown in Fig. 4.4. Each scatter plot is annotated with the corresponding RMSE. The results indicate that when the correlation threshold is set to 0.3—selecting all features with moderate or higher correlation—the performance of feature selection is optimal, with the RMSE significantly lower than that of predictions made without feature selection (threshold = 0.0) and those made using only AQI. This demonstrates that at this threshold, neither too many relevant features were discarded nor too many weakly correlated features were retained.

4.3.3 *Wrapper Methods*

4.3.3.1 Theoretical Basis

Wrapper methods are a feature selection technique that evaluates the performance of a predictive model using different subsets of features. Unlike filter methods, which rank features based on statistical measures independent of the model, wrapper methods integrate the feature selection process directly into model training and evaluation. The core idea of wrapper methods is to repeatedly train the model on various feature subsets and select the one that yields the best model performance.

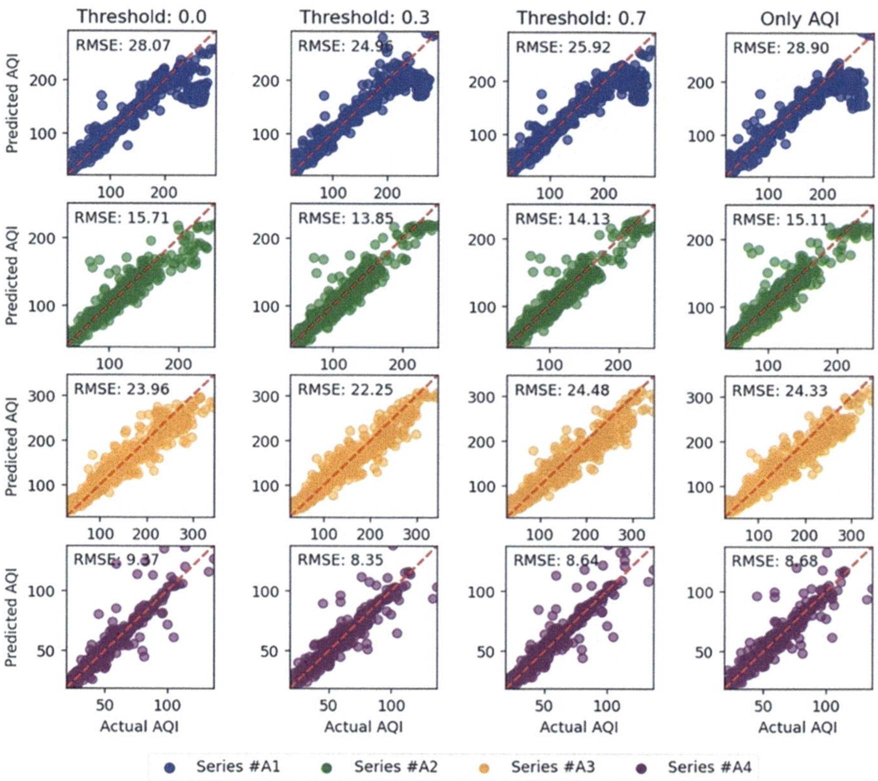

Fig. 4.4 The scatter plots of predicted AQI versus actual AQI

The key advantage of wrapper methods is their close interaction with the model, which allows them to select the most suitable features for a specific algorithm. However, the computational cost is relatively high because every feature subset needs to be evaluated by training the model. This can be resource-intensive, particularly when dealing with large-scale or high-dimensional datasets.

Wrapper methods can generally be categorized into three main approaches (Wah et al. 2018):

- Forward selection: starts with an empty set of features and gradually adds the feature that most improves model performance until no further improvement can be made.
- Backward elimination: starts with the complete set of features and iteratively removes the least significant feature until performance begins to degrade.

- Recursive feature elimination (RFE): a popular wrapper method where the model is trained repeatedly, removing the least important features in each iteration until a desired number of features is reached.

At the heart of wrapper methods is an optimization problem, where the objective is to select the optimal feature subset $S \subseteq F$that maximizes the model's predictive performance. Here, F represents the full set of features, and S is a subset of those features. Given a loss function L (such as mean squared error or log loss), the goal is to minimize the model's loss on the training set. The symbol * represents the meaning of "best" or "optimal".

$$S^{*} = \arg\min_{S \subseteq F} L\left(f\left(X_{S}\right), y\right) \tag{4.5}$$

where $f(X_S)$ is the model trained on the feature subset S, y is the corresponding true label or target value, represents the loss function, $L(f(X_S), y)$ measuring the discrepancy between the predicted and actual values.

Specific algorithms for wrapper methods can be implemented through the following steps:

4.4 Forward Selection

Initialization: Start with an empty feature set, $S_0 = \phi$.

Iteration: At each step, select the feature $x_i \in F - S$ that, when added to S, minimizes the loss function $L\left(f\left(X_{S \cup \{x_i\}}\right), y\right)$.

$$x_i^{*} = \arg\min_{x_i \in F - S} L\left(f\left(X_{S \cup \{x_i\}}\right), y\right) \tag{4.6}$$

Update: Add the selected feature to the set, $S = S \cup \left\{x_i^{*}\right\}$.

Stopping Condition: Stop when adding new features no longer significantly improves performance.

4.5 Backward Elimination

Initialization: Start with the full set of features, $S_0 = F$.

Iteration: At each step, select the feature $x_i \in S$ that causes the smallest increase in the loss function when removed:

$$x_i^{*} = \arg\min_{x_i \in F - S} L\left(f\left(X_{S - \{x_i\}}\right), y\right) \tag{4.7}$$

Update: Remove the selected feature from the set, $S = S - \left\{x_i^{*}\right\}$.

Stopping Condition: Stop when removing any more features causes a significant degradation in model performance.

4.6 Recursive Feature Elimination (RFE)

Initialization: Begin with the full set of features, $S_0 = F$.

Iteration: Train the model and rank features based on their importance (such as the absolute values of their coefficients). In each iteration, remove the least important feature, retraining the model with the remaining features:

$$x_i^* = \arg\min_{x_i \in S} \text{FeatureImportance}\left(f\left(X_S\right)\right) \tag{4.8}$$

Update: Remove the least important feature from the set, $S = S - \left\{x_i^*\right\}$.

Stopping Condition: Stop when the desired number of features is reached.

4.6.1 Modeling Step

The experiment for the wrapper method was conducted using RFE technique to evaluate feature importance. The training model used in this experiment is XGBoost. The procedure is as follows: first, XGBoost is applied to the training set for 1-steahead prediction fitting. Then, based on the feature importance ranking derived from the fitted model, RFE is performed to assess prediction errors with different numbers of retained features.

It is important to note that, given a time window of 24 and each sample containing 7 features, XGBoost requires features to be flattened into a one-dimensional input, unlike LSTM, which can directly accept time series data. Thus, the model processes 7 × 24 input values, with each value corresponding to a specific importance score. Using data from Beijing as an example, the importance distribution of XGBoost for different times and features is shown in Fig. 4.5. The time axis in the figure represents the forward sequence within the time window. The analysis reveals that in the 1-step AQI prediction (predicting AQI at time $x + 1$), the AQI values at time x and $x-23$ exhibit significantly higher importance compared to other input features. Furthermore, when comparing feature importance across all time points, the values at time x and $x-23$ consistently rank higher than those between $x-1$ and $x-22$. This indicates that for 1-step prediction, both the most recent value (1 h prior) and the value from 24 h prior play crucial roles, highlighting the strong diurnal cyclicity of AQI patterns.

Based on the importance ranking, the top 5, 10, 30, 100, and 168 inputs (i.e., all inputs) were retained, while the remaining unselected inputs were filled with zeros to ensure consistent data dimensions. Subsequently, an LSTM model was used for time series prediction of AQI. The scatter plots in Fig. 4.6 compare the predicted

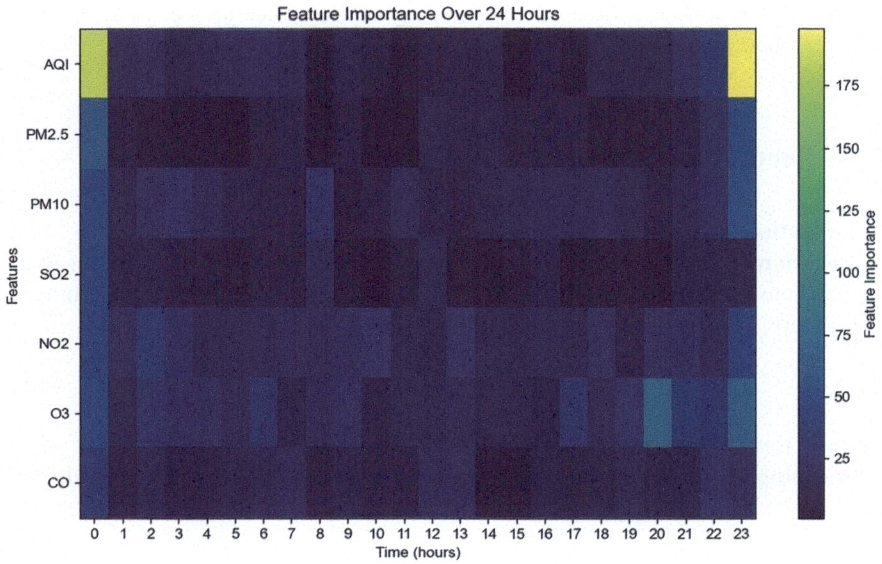

Fig. 4.5 The feature importance distribution in XGBoost

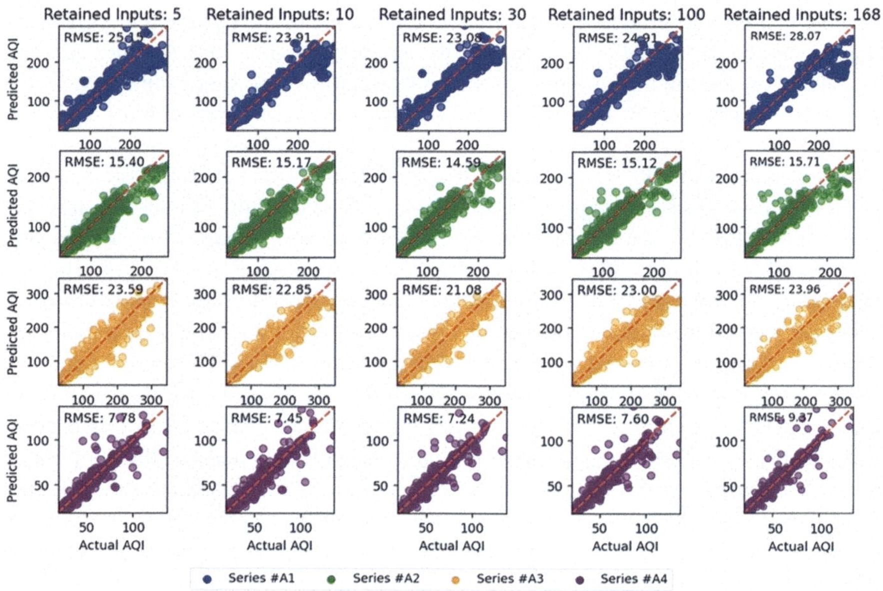

Fig. 4.6 The scatter plots of predicted AQI versus actual AQI

AQI values with the actual AQI values. It is evident that predictive performance improves when the top 30 inputs are retained. Therefore, for the subsequent performance comparisons, only the top 30 important inputs will be utilized.

4.6.2 Embedded Methods

Embedded methods are a class of techniques that tightly integrate feature selection with model training. In these methods, feature selection is not a separate process from model training but is embedded as part of the model itself (Jović et al. 2015). During training, the model simultaneously optimizes its parameters and automatically selects the most relevant features. Through the model's internal mechanisms, embedded methods can effectively assess the importance of features and select those most beneficial for prediction. This approach often improves the model's predictive accuracy and stability, especially when handling high-dimensional data.

Typical models that employ embedded methods include decision trees, Lasso regression, random forests, and XGBoost (Jović et al. 2015). In these models, feature selection is accomplished through intrinsic mechanisms. For instance, Lasso regression applies L1 regularization to shrink or eliminate the coefficients of less important features, while decision trees and random forests use criteria like information gain or Gini index to evaluate feature importance. In XGBoost, feature selection is automatically performed through the boosting mechanism, selecting the most critical features during model training. The importance distribution of inputs in XGBoost is shown in Fig. 4.5. These methods not only achieve feature selection but also optimize overall model performance.

In contrast, filter methods and wrapper methods typically function as steps independent of model training. Filter methods rely on statistical measures or feature scores for selection and do not depend on any specific model. Wrapper methods, on the other hand, use iterative model training to select features. Embedded methods differ from both, as they include feature selection as an integral part of the training process itself (Chen and Guestrin 2016). As such, embedded methods cannot be directly compared with filter or wrapper methods since they are not separate steps that occur before or after training but are instead embedded within the training process itself.

4.7 Feature Extraction of Air Quality Data

4.7.1 Feature Extraction Performance Evaluation

For ease of horizontal comparison, the data used in this section is the same as in Sect. 4.2, labeled as series #A1, A2, A3, and A4, with the performance evaluation metrics remaining unchanged.

4.7.2 Statistical Feature Extraction

4.7.2.1 Theoretical Basis

Statistical feature extraction involves summarizing and representing raw data through various statistical measures, capturing key patterns and distributions (Fan et al. 2024). This method is grounded in statistical theory, which assumes that the underlying structure of the data can be described and interpreted using mathematical summaries. Commonly used features include measures of central tendency, dispersion, and shape, such as mean, variance, standard deviation, skewness, and kurtosis. These features offer insight into the overall behavior of the dataset and are particularly useful in revealing trends, periodicity, or anomalies in time series data, such as air quality measurements.

Transforming raw time series data into statistical summaries facilitates the extraction of hidden features, preserving essential information about the variability and characteristics of the data (Fan et al. 2024). This allows for more efficient training and prediction using machine learning models, such as LSTM or other advanced methods, facilitating more effective and interpretable results. In air quality monitoring, the extraction of statistical features from pollutant concentrations like AQI, PM2.5, and others provides a clearer understanding of environmental patterns and helps in developing accurate forecasting models.

In this part, five statistics including moving average, rate of change, standard deviation, skewness, and kurtosis were used for feature extraction. The calculation formulas are as follows:

$$\mathrm{MA}(t) = \frac{1}{N} \sum_{i=t-N+1}^{t} x_i \tag{4.9}$$

$$\mathrm{ROC}(t) = \frac{x_t - x_{t-N}}{x_{t-N}} \times 100\% \tag{4.10}$$

$$\mathrm{StdDev}(t) = \sqrt{\frac{1}{N} \sum_{i=t-N+1}^{t} (x_i - \mu)^2} \tag{4.11}$$

$$\mathrm{StdDev}(t) = \sqrt{\frac{1}{N} \sum_{i=t-N+1}^{t} (x_i - \mu)^2} \tag{4.12}$$

$$\mathrm{Kurtosis}(t) = \frac{\frac{1}{N} \sum_{i=t-N+1}^{t} (x_i - \mu)^4}{\left(\frac{1}{N} \sum_{i=t-N+1}^{t} (x_i - \mu)^2 \right)^2} - 3 \tag{4.13}$$

where t indicates the current time, N is the window size, x_t is the value of the current time, x_{t-N} is the value N time steps before, and μ is the mean value in the window.

4.7.2.2 Modeling Step

By applying the sliding window method, the statistical analysis of feature values within each time window provides the distribution of features along the time axis. Figure 4.7 illustrates the statistical value series of AQI in series #A1, including moving average, rate of change, standard deviation (SD), skewness, and kurtosis. Statistical value series were extracted for all 7 features, expanding the input to 7×6 features. These were then used as input for LSTM to perform time series prediction, with the scatter plots shown in Fig. 4.8.

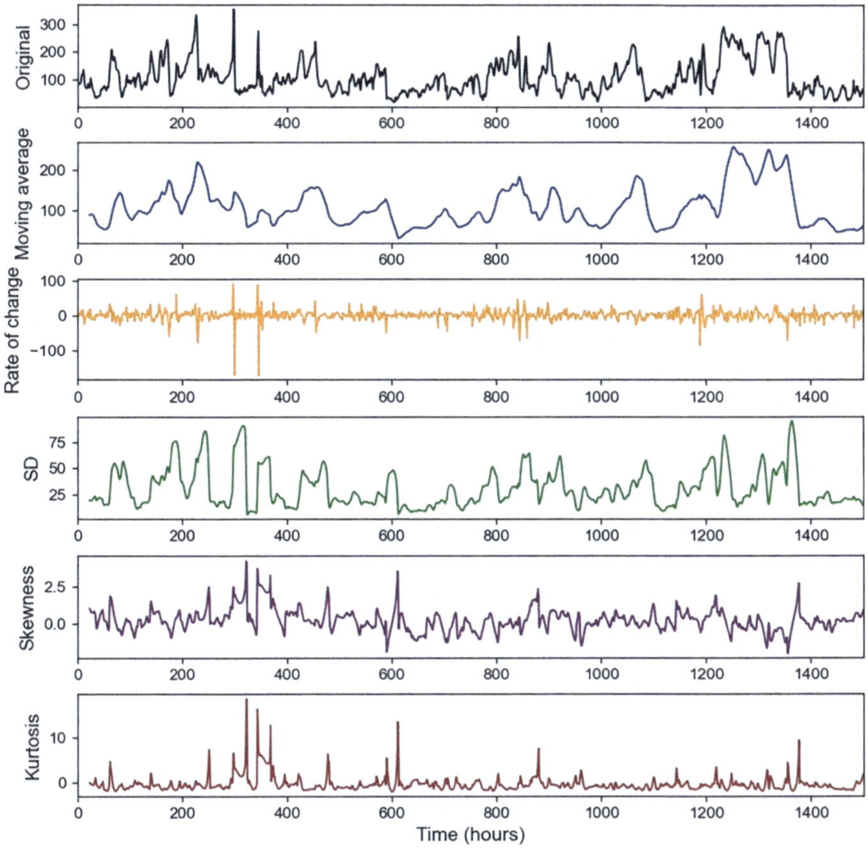

Fig. 4.7 Statistical value series of AQI in series #A1

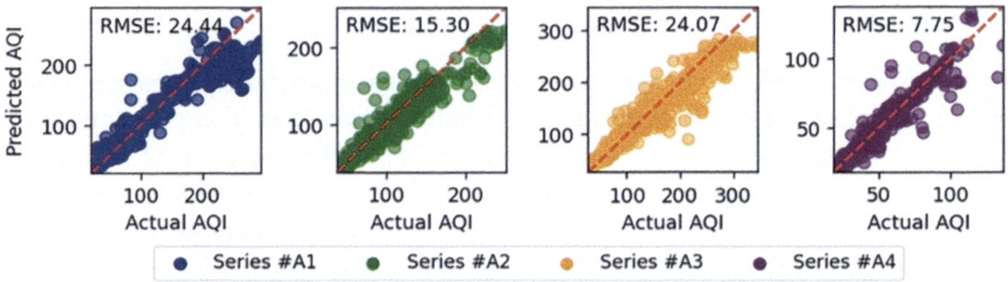

Fig. 4.8 The scatter plots of predicted AQI versus actual AQI

4.7.3 Time-Frequency Analysis

4.7.3.1 Theoretical Basis

Time-frequency analysis is a method used to examine the signal in both time and frequency domains simultaneously (Pachori 2023). This approach is particularly useful for non-stationary signals, where the frequency content changes over time, such as in air quality data affected by periodic events like diurnal cycles, meteorological conditions, and human activities.

The theoretical foundation of time-frequency analysis lies in the joint representation of signals in both the time and frequency domains, allowing for the observation of how the spectral content of a signal evolves over time. One of the most widely used techniques in this category is the short-time Fourier transform (STFT), which divides the signal into shorter segments and applies the Fourier transform to each segment, thus providing a time-localized frequency spectrum (Gao et al. 2015). However, due to the fixed window size used in STFT, it can face limitations when dealing with signals that require both high time and frequency resolution.

To overcome this, more advanced techniques such as the wavelet transform are employed (Ahmed et al. 2021). The wavelet transform uses varying window sizes, offering better adaptability for signals with both slow and fast-changing frequency components. It decomposes the signal into components at different scales and positions, giving a more flexible time-frequency resolution compared to STFT (Grobbelaar et al. 2022).

Wavelet transform decomposes a signal layer by layer, resulting in signal components at different scales (or resolutions). Each layer is divided into two parts: approximation coefficients (A) and detail coefficients (D), representing low-frequency and high-frequency information, respectively. Wavelet transforms can be classified into two types: continuous wavelet transform (CWT) and discrete wavelet transform (DWT). DWT uses specific scaling and translation parameters, analyzing the signal with discrete scales and time shifts. With higher computational efficiency, DWT is commonly used in practical applications (Khorrami and Moavenian 2010). Therefore, this section employs DWT for the experiment. Through stepwise decomposition, DWT represents the signal as a combination of approximation and detail coefficients at different scales.

$$x(t)=\sum_{k}A_{j_0,k}\varphi_{j_0,k}(t)+\sum_{j=j_0}^{J-1}\sum_{k}D_{j,k}\psi_{j,k}(t) \tag{4.14}$$

where, $\varphi_{j_0,k}(t)$ is the scaling function used to represent the low-frequency approximation components; $\psi_{j,\,k}(t)$ is the wavelet function used to represent the high-frequency detail components; $A_{j_0,k}$ are the approximation coefficients capturing the low-frequency part of the signal; $D_{j,\,k}$ are the detail coefficients capturing the high-frequency part of the signal.

By continuously applying DWT,. a signal can be decomposed into multiple levels. The low-frequency component at each level can undergo further wavelet transformation, generating new approximation and detail coefficients, forming a pyramid structure. By continuously applying DWT, a signal can be decomposed into multiple levels. The low-frequency component at each level can undergo further wavelet transformation, generating new approximation and detail coefficients, forming a pyramid structure.

$$A_{j+1}[k]=\sum_{n}A_j[n]\cdot g[2k-n] \tag{4.15}$$

$$D_{j+1}[k]=\sum_{n}A_j[n]\cdot h[2k-n] \tag{4.16}$$

where j refers to the decomposition of several layers, $g[n]$ is a low-pass filter used to calculate the approximation coefficient, $h[n]$ is a high-pass filter used to calculate the detail coefficient, $A_1[k]=\sum_{n}x[n]\cdot g[2k-n]$, and $D_1[k]=\sum_{n}x[n]\cdot h[2k-n]$.

After multiple levels of decomposition, the signal can be represented as:

$$x(t)\rightarrow\{A_j,D_j,D_{j-1},\ldots,D_1\} \tag{4.17}$$

4.7.3.2 Modeling Step

First, DWT is applied to decompose all features. Figure 4.9 illustrates the components obtained after applying a 5-level DWT on the AQI feature in series #A1. According to Eqs. (4.15) and (4.16), each decomposition step reduces the sequence length by half, meaning that the length of the components decreases as the decomposition level increases. Before inputting these components into the LSTM model, to ensure that all components have the same length, we need to apply zero-padding at regular intervals. This ensures data consistency and allows the model's input to be processed smoothly. Additionally, to determine the optimal number of decomposition levels in DWT, we conduct a comparative evaluation, considering five different levels: 2, 3, 4, 5, and 6. This comparison not only helps to understand the impact of different decomposition levels on model performance but also ensures that the optimal level is selected, thereby enhancing the predictive performance of the subsequent LSTM model. The final results, as shown in Fig. 4.10, indicate that a decomposition level of 5 yields the best outcome.

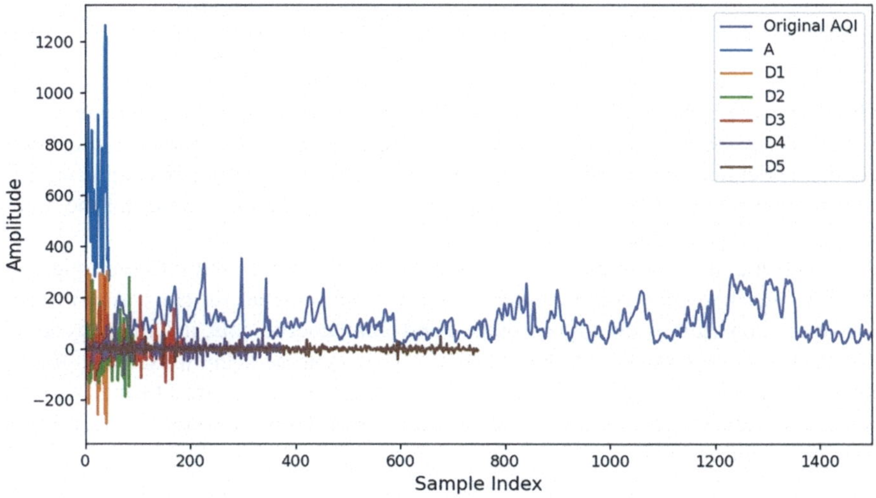

Fig. 4.9 Components after 5-Level DWT on AQI

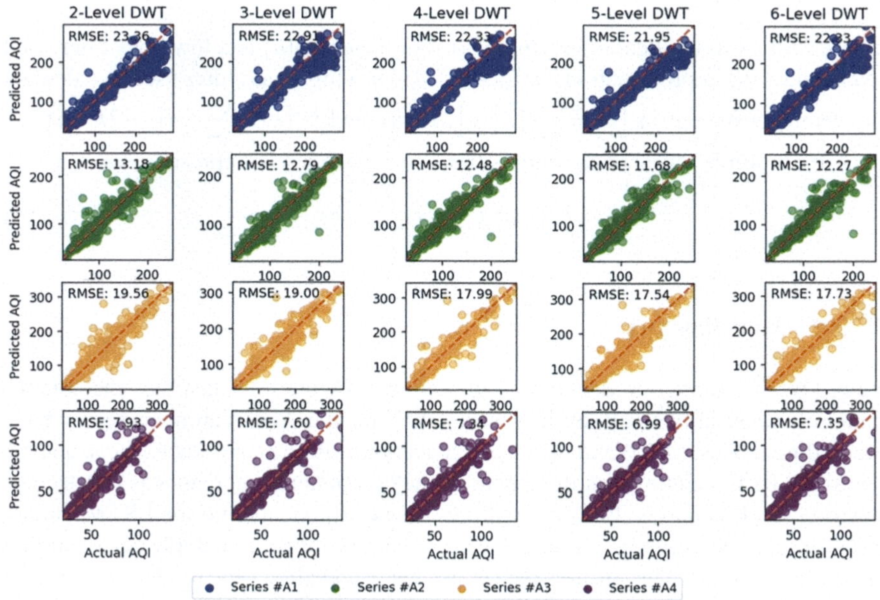

Fig. 4.10 The scatter plots of predicted AQI versus actual AQI

4.8 Identification Performance Comparison

4.8.1 Performance Comparison of Feature Selection

In this chapter, the filter method and wrapper method are both applied for feature selection on multivariate time-series data. Since embedded methods are inherently integrated within specific models, they are not directly comparable to the filter and wrapper methods, so no performance comparison is made here. To illustrate, evaluating the performance using 1-step, 2-step, and 3-step predictions, with the corresponding metrics shown in Table 4.2. Among the results, for each forecast step and across different series, the best-performing metric for each case is highlighted in bold. The comparison of RMSE across different methods and prediction steps is shown in Fig. 4.11. The RMSE box plot of each series in 1-step forecast is shown in Fig. 4.12.

From Table 4.2, Figs. 4.11 and 4.12, it can be concluded as follows:

1. The wrapper method consistently exhibits superior overall performance across all prediction steps. For example, in the case of series #A1, the wrapper method achieved an MAE of 14.50 and an RMSE of 23.08, outperforming both the filter method and the unprocessed condition. This superior performance indicates that the wrapper method, which utilizes the predictive model to evaluate the importance of features, effectively enhances the model's ability to capture complex relationships within the data. The results suggest that the iterative process of selecting and evaluating features based on model performance can lead to more accurate predictions, particularly in dynamic environments like air quality monitoring.
2. As the prediction horizon extends from 1-step to 3-step forecasts, a general deterioration in performance metrics such as MAE, RMSE, P, and KGE is observed for all three methods. For instance, the MAE for series #A1 increased from 15.35 in the 1-step prediction to 33.76 in the 3-step prediction using the filter method. However, the wrapper method still maintained a higher level of accuracy compared to the other two methods, underscoring its robustness in handling longer-term predictions. This trend highlights the inherent challenges in time series forecasting, where the ability to predict future values diminishes as the time interval increases. The wrapper method's relative advantage in maintaining accuracy suggests that it may be better equipped to navigate these challenges through effective feature selection.
3. The unprocessed condition, which incorporates all features without any feature selection, consistently yields the lowest accuracy metrics across the board. For example, in the 3-step prediction for series #A1, the unprocessed condition recorded a KGE of only 0.5321, indicating significant room for improvement. This finding emphasizes the critical role that effective feature selection plays in enhancing model performance in time series forecasting. By reducing the dimensionality of the input space and eliminating irrelevant or redundant fea-

Table 4.2 The feature selection performance of the filter method and wrapper method

Prediction steps	Method	Series	Accuracy performance			
			MAE	RMSE	P	KGE
1-step	Filter method	#A1	15.35	24.96	0.9601	0.8242
		#A2	**10.91**	**13.85**	**0.9430**	**0.8722**
		#A3	15.83	22.25	0.9584	0.8824
		#A4	4.45	8.35	0.9431	0.9310
	Wrapper method	#A1	**14.50**	**23.08**	**0.9638**	**0.8389**
		#A2	11.07	14.59	0.9419	0.8670
		#A3	**15.02**	**21.08**	**0.9627**	**0.8966**
		#A4	**3.66**	**7.24**	**0.9468**	**0.9439**
	Unprocessed	#A1	16.61	28.07	0.9520	0.8160
		#A2	12.56	15.71	0.9275	0.8479
		#A3	16.28	23.96	0.9571	0.8756
		#A4	5.01	9.37	0.9387	0.9204
2-step	Filter method	#A1	23.46	35.86	0.8941	0.7442
		#A2	17.59	23.56	0.8624	0.7624
		#A3	26.34	37.95	0.8698	0.7358
		#A4	5.48	9.10	0.9179	0.9058
	Wrapper method	#A1	**23.27**	**34.41**	**0.9022**	**0.7698**
		#A2	**16.71**	**22.05**	**0.8713**	**0.7726**
		#A3	**25.94**	**36.65**	**0.8869**	**0.7418**
		#A4	**5.17**	**8.88**	**0.9209**	**0.9146**
	Unprocessed	#A1	24.39	37.17	0.8891	0.6942
		#A2	17.93	25.40	0.8594	0.7533
		#A3	27.37	40.56	0.8589	0.7215
		#A4	6.16	9.81	0.9034	0.9003
3-step	Filter method	#A1	33.76	51.35	0.7879	0.5387
		#A2	**20.46**	**27.46**	**0.7923**	**0.6645**
		#A3	33.72	43.84	0.8249	0.6606
		#A4	6.74	11.36	0.8747	0.8687
	Wrapper method	#A1	**33.07**	**50.58**	**0.7914**	**0.5483**
		#A2	20.83	27.98	0.7865	0.6600
		#A3	**32.54**	**42.53**	**0.8396**	**0.6688**
		#A4	**6.27**	**10.29**	**0.8949**	**0.8881**
	Unprocessed	#A1	35.13	53.15	0.7644	0.5321
		#A2	23.57	30.38	0.7547	0.6402
		#A3	35.23	45.92	0.8148	0.6584
		#A4	7.19	12.02	0.8583	0.8430

tures, the filter and wrapper methods allow the model to focus on the most influential variables, thereby reducing noise and improving prediction accuracy. The stark contrast in performance between the unprocessed condition and the feature selection techniques reinforces the necessity of strategic feature selection in achieving reliable forecasting outcomes in complex datasets.

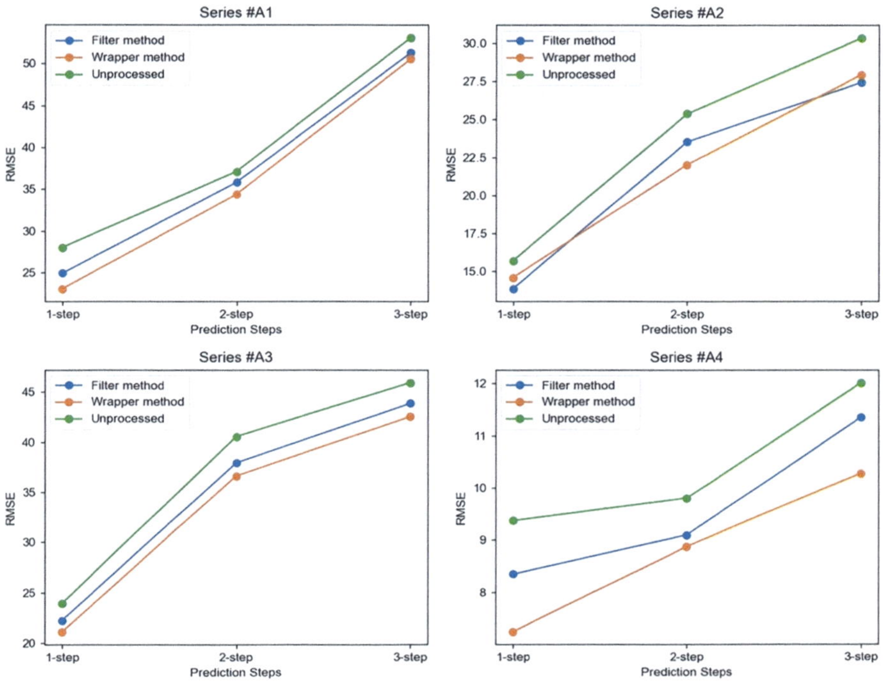

Fig. 4.11 Comparison of RMSE across different methods

4. Although the wrapper method demonstrates superior overall performance, the filter method can also achieve optimal feature selection outcomes in certain situations. This observation highlights the importance of conducting targeted evaluations based on the specific characteristics of different datasets, rather than mechanically applying a fixed method. In particular, the Filter method's efficiency in scenarios with simpler relationships or when computational resources are limited suggests that it can be a viable alternative or complement to the Wrapper method. Therefore, practitioners should carefully consider the context and requirements of their analyses to select the most appropriate feature selection technique, ensuring that the chosen method aligns with the unique attributes of the dataset at hand.
5. The boxplot analysis further validates the stability and reliability of the obtained results, demonstrating that these outcomes are not due to random chance. By comparing the RMSE distributions across different methods, it is clearly shown that the application of the filter method and wrapper method preprocessing techniques significantly reduces the median values, indicating a notable improvement in predictive performance. Consequently, it can be inferred that these preprocessing methods play a substantial role in enhancing the model's predictive accuracy, effectively optimizing overall model performance and leading to more accurate forecasts.

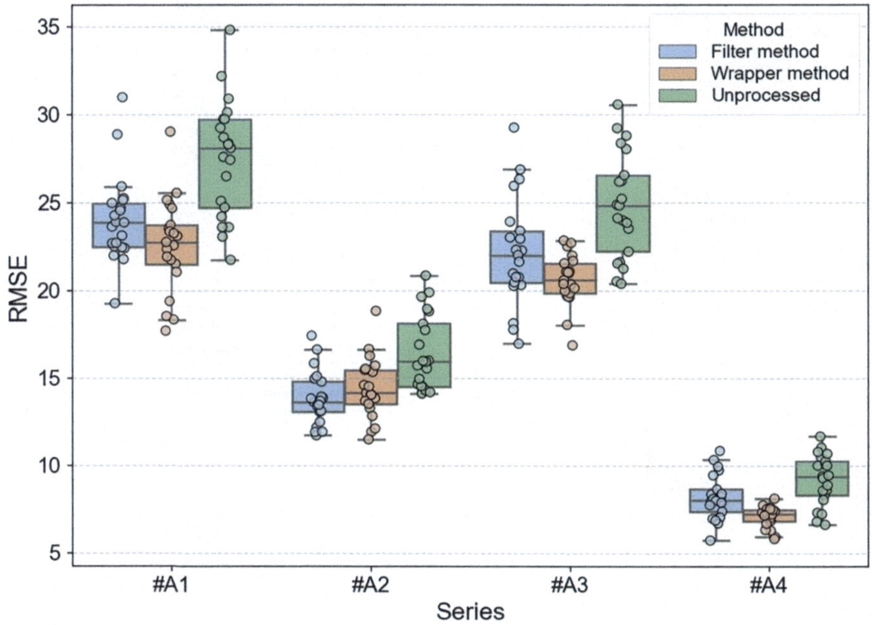

Fig. 4.12 1-step RMSE comparison across methods and series

4.8.2 Performance Comparison of Feature Extraction

In this section, the performance of statistical feature extraction and time-frequency analysis for multivariate time-series data is evaluated. These two methods offer complementary approaches to feature extraction, each with its own strengths. For statistical feature extraction, a range of statistical metrics, such as mean, variance, skewness, and kurtosis, are applied to capture the fundamental characteristics of the time series. This approach helps provide a comprehensive understanding of data distribution and variability, which is critical for subsequent modeling tasks. On the other hand, time-frequency analysis utilizes the DWT to decompose the time series into different frequency components, aiding in the identification of hidden temporal patterns and dynamic changes within the data. In Table 4.3, we summarize the results for 1-step, 2-step, and 3-step predictions, with the best-performing metrics highlighted in bold, emphasizing the optimal statistical feature extraction under different forecasting scenarios. Figure 4.13 presents the performance comparison across different methods and prediction steps, providing a visual overview of each method's contribution to predictive accuracy. The RMSE box plot of each series in 1-step forecast is shown in Fig. 4.14.

From Table 4.3, Figs. 4.13 and 4.14, it can be concluded as follows:

Table 4.3 The feature extraction performance of the statistical analysis and DWT

Prediction steps	Method	Series	Accuracy performance			
			MAE	RMSE	P	KGE
1-step	Statistical analysis	#A1	15.19	25.29	0.9601	0.8202
		#A2	10.38	14.50	0.9444	0.8660
		#A3	15.46	22.46	0.9609	0.8841
		#A4	4.63	8.52	0.9445	0.9302
	DWT	#A1	**14.07**	**21.95**	**0.9695**	**0.8270**
		#A2	**8.27**	**11.68**	**0.9541**	**0.8846**
		#A3	**12.24**	**17.54**	**0.9728**	**0.9197**
		#A4	**3.65**	**6.99**	**0.9513**	**0.9502**
	Unprocessed	#A1	16.61	28.07	0.9520	0.8160
		#A2	12.56	15.71	0.9275	0.8479
		#A3	16.28	23.96	0.9571	0.8756
		#A4	5.01	9.37	0.9387	0.9204
2-step	Statistical analysis	#A1	23.31	35.83	0.8841	0.7493
		#A2	16.29	23.74	0.8673	0.7756
		#A3	25.93	37.81	0.8774	0.7437
		#A4	5.87	9.01	0.9092	0.9124
	DWT	#A1	**21.19**	**31.03**	**0.9173**	**0.7974**
		#A2	**14.83**	**19.27**	**0.8896**	**0.8103**
		#A3	**24.69**	**34.98**	**0.8953**	**0.7647**
		#A4	**4.38**	**7.45**	**0.9275**	**0.9321**
	Unprocessed	#A1	24.39	37.17	0.8891	0.6942
		#A2	17.93	25.40	0.8594	0.7533
		#A3	27.37	40.56	0.8589	0.7215
		#A4	6.16	9.81	0.9034	0.9003
3-step	Statistical analysis	#A1	33.62	51.83	0.7816	0.5574
		#A2	22.09	28.46	0.7611	0.6526
		#A3	33.74	43.78	0.8283	0.6701
		#A4	6.56	10.85	0.8688	0.8615
	DWT	#A1	**30.13**	**46.63**	**0.8152**	**0.5945**
		#A2	**18.74**	**25.53**	**0.7974**	**0.6844**
		#A3	**29.46**	**38.94**	**0.8551**	**0.6902**
		#A4	**5.84**	**8.86**	**0.9007**	**0.9019**
	Unprocessed	#A1	35.13	53.15	0.7644	0.5321
		#A2	23.57	30.38	0.7547	0.6402
		#A3	35.23	45.92	0.8148	0.6584
		#A4	7.19	12.02	0.8583	0.8430

1. The DWT consistently achieves superior performance compared to statistical analysis and unprocessed data across 1-step, 2-step, and 3-step predictions. In all series, DWT shows the lowest MAE and RMSE, alongside higher Pearson's correlation and KGE scores. This indicates that DWT is more effective at capturing

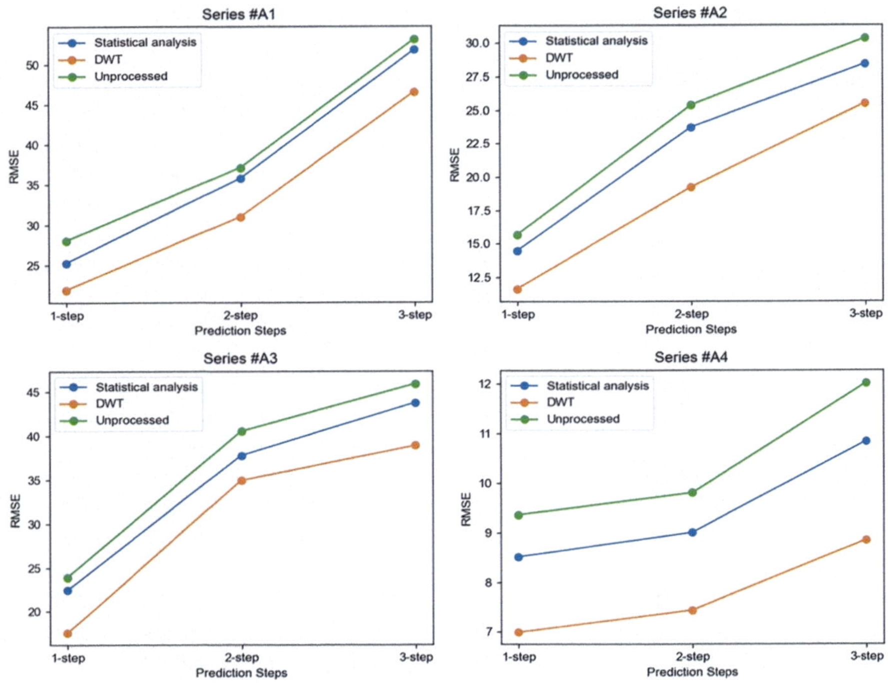

Fig. 4.13 Comparison of RMSE across different feature extraction methods

underlying time-series patterns, providing a more robust input structure for the LSTM model, which results in more accurate AQI predictions.

2. The results clearly highlight the importance of proper feature extraction for time-series forecasting. DWT, through its decomposition of signals, isolates key components that improve model inputs, leading to enhanced prediction accuracy. In contrast, the unprocessed data, lacking any form of feature extraction, consistently performs the worst, particularly in multi-step predictions. Statistical analysis, while better than unprocessed data, still falls short of DWT in capturing the complex temporal patterns needed for high-quality AQI forecasting.
3. Similarly, the boxplot analysis confirms the stability and reliability of the obtained data, ruling out the possibility of random occurrence. Notably, after applying the DWT preprocessing technique, the RMSE distribution became more compact, with a significant reduction in the median values, indicating an improvement in the model's predictive accuracy. This suggests that DWT successfully extracted patterns and trends from the data features. Overall, the preprocessed models exhibited reduced errors across multiple prediction steps, highlighting the advantage of preprocessing in enhancing predictive accuracy.

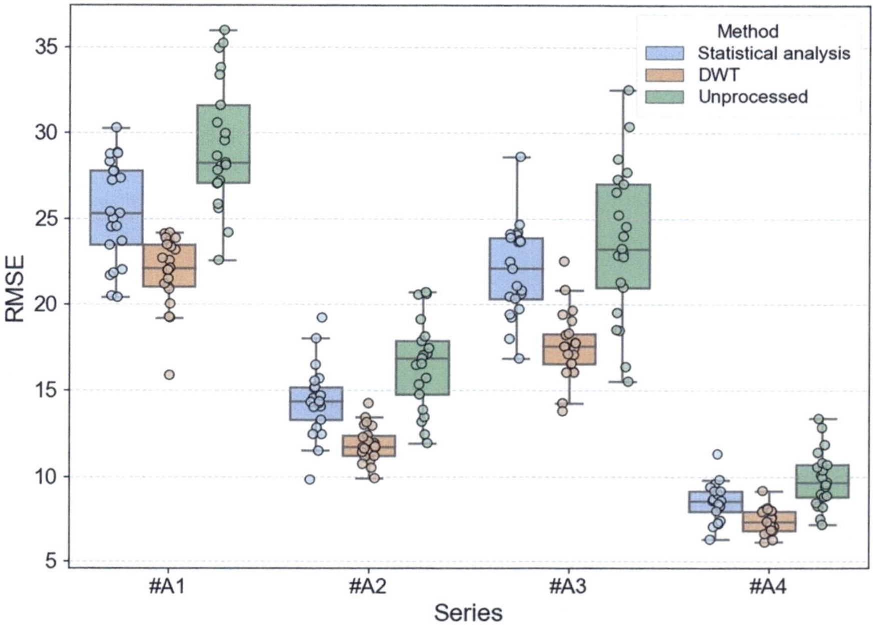

Fig. 4.14 1-step RMSE comparison across methods and series

This provides strong evidence of the effectiveness of these preprocessing methods and reinforces their role in optimizing model performance.

4.9 Conclusions

In this chapter, we examined the critical role of data identification in enhancing the predictive performance of air quality monitoring models. The wrapper method consistently demonstrated superior accuracy across prediction steps through an in-depth performance comparison, particularly in managing complex and dynamic time series data. This advantage highlights its capability to handle extended prediction horizons where forecasting precision typically declines. Despite this, with its computational efficiency, the filter method can be effective in simpler scenarios or resource-limited settings. Furthermore, the performance contrast between processed and unprocessed data conditions reaffirms the value of strategic feature selection in improving prediction reliability by eliminating noise and focusing on impactful variables.

For feature extraction, the comparison between statistical feature extraction and time-frequency analysis methods showcased their complementary strengths.

Statistical metrics captured core data distributions, while DWT-based time-frequency analysis provided insights into underlying temporal patterns. The boxplot analysis validated the stability of these results, confirming that the preprocessing methods significantly enhance model performance and lead to more accurate forecasts. This chapter's findings underscore the necessity of selecting suitable data identification techniques to tailor air quality monitoring models to specific dataset characteristics, thus enabling more effective environmental management.

References

Afifa, Arshad K, Hussain N, Ashraf MH, Saleem MZ (2024) Air pollution and climate change as grand challenges to sustainability. Sci Total Environ 928:172370

Ahmed S, Frikha M, Hussein TDH, Rahebi J (2021) Optimum feature selection with particle swarm optimization to face recognition system using Gabor wavelet transform and deep learning. Biomed Res Int 2021(1):6621540

Austin E, Coull B, Thomas D, Koutrakis P (2012) A framework for identifying distinct multipollutant profiles in air pollution data. Environ Int 45:112–121

Baldasano JM, Valera E, Jiménez P (2003) Air quality data from large cities. Sci Total Environ 307(1–3):141–165

Belsley DA, Kuh E, Welsch RE (2005) Regression diagnostics: identifying influential data and sources of collinearity. Wiley, Hoboken

Breuer D (1999) Monitoring ambient air quality for health impact assessment, vol 85. WHO Regional Office Europe

Carslaw DC, Ropkins K (2012) Openair—an R package for air quality data analysis. Environ Model Softw 27:52–61

Chen T, Guestrin C (2016) Xgboost: a scalable tree boosting system. In: Proceedings of the 22nd ACM SIGKDD international conference on knowledge discovery and data mining, pp 785–794

El Aboudi N, Benhlima L (2016) Review on wrapper feature selection approaches. In: 2016 international conference on engineering & MIS (ICEMIS). IEEE, pp 1–5

Fan GF, Han YY, Li JW, Peng LL, Yeh YH, Hong WC (2024) A hybrid model for deep learning short-term power load forecasting based on feature extraction statistics techniques. Expert Syst Appl 238:122012

Gao H, Liang L, Chen X, Xu G (2015) Feature extraction and recognition for rolling element bearing fault utilizing short-time Fourier transform and non-negative matrix factorization. Chin J Mech Eng 28:96–105

Gong H, Li Y, Zhang J, Zhang B, Wang X (2024) A new filter feature selection algorithm for classification task by ensembling Pearson correlation coefficient and mutual information. Eng Appl Artif Intell 131:107865

Grobbelaar M, Phadikar S, Ghaderpour E, Struck AF, Sinha N, Ghosh R, Ahmed MZI (2022) A survey on denoising techniques of electroencephalogram signals using wavelet transform. Signals 3(3):577–586

Hasan R, Chu C (2022) Noise in datasets: what are the impacts on classification performance? [Noise in datasets: what are the impacts on classification performance?]. In: Proceedings of the 11th international conference on pattern recognition applications and methods

Jović A, Brkić K, Bogunović N (2015) A review of feature selection methods with applications. In: 2015 38th international convention on information and communication technology, electronics and microelectronics (MIPRO). IEEE, pp 1200–1205

Khorrami H, Moavenian M (2010) A comparative study of DWT, CWT and DCT transformations in ECG arrhythmias classification. Expert Syst Appl 37(8):5751–5757

Lal TN, Chapelle O, Weston J, Elisseeff A (2006) Embedded methods. In: Feature extraction: foundations and applications. Springer, pp 137–165

Li J, Cheng K, Wang S, Morstatter F, Trevino RP, Tang J, Liu H (2017) Feature selection: a data perspective. ACM Comput Surv (CSUR) 50(6):1–45

Luo X, Jiang R, Yang B, Qin H, Hu H (2024) Air quality visualization analysis based on multivariate time series data feature extraction. J Vis 27:567–584

Morabito FC, Versaci M (2003) Fuzzy neural identification and forecasting techniques to process experimental urban air pollution data. Neural Netw 16(3–4):493–506

Mutlag WK, Ali SK, Aydam ZM, Taher BH (2020) Feature extraction methods: a review. J Phys Conf Ser 1591:012028

Pachori RB (2023) Time-frequency analysis techniques and their applications. CRC Press, Boca Raton

Rabie R, Asghari M, Nosrati H, Niri ME, Karimi S (2024) Spatially resolved air quality index prediction in megacities with a CNN-Bi-LSTM hybrid framework. Sustain Cities Soc 109:105537

Ratner B (2009) The correlation coefficient: its values range between +1/−1, or do they? J Target Meas Anal Mark 17(2):139–142

Singh D, Dahiya M, Kumar R, Nanda C (2021) Sensors and systems for air quality assessment monitoring and management: a review. J Environ Manag 289:112510

Thudumu S, Branch P, Jin J, Singh J (2020) A comprehensive survey of anomaly detection techniques for high dimensional big data. J Big Data 7:1–30

Tian Y, Yao XA, Mu L, Fan Q, Liu Y (2020) Integrating meteorological factors for better understanding of the urban form-air quality relationship. Landsc Ecol 35:2357–2373

Wah YB, Ibrahim N, Hamid HA, Abdul-Rahman S, Fong S (2018) Feature selection methods: case of filter and wrapper approaches for maximising classification accuracy. Pertanika J Sci Technol 26(1):329–340

Yang Z, Wang J (2017) A new air quality monitoring and early warning system: air quality assessment and air pollutant concentration prediction. Environ Res 158:105–117

Yu Y, Si X, Hu C, Zhang J (2019) A review of recurrent neural networks: LSTM cells and network architectures. Neural Comput 31(7):1235–1270

Zebari R, Abdulazeez A, Zeebaree D, Zebari D, Saeed J (2020) A comprehensive review of dimensionality reduction techniques for feature selection and feature extraction. J Appl Sci Technol Trends 1(1):56–70

Chapter 5
Data Preprocessing in Air Quality Monitoring

Abstract Efficient air quality monitoring relies on reliable data preprocessing to ensure the precision and stability of predictive models. In urban environments, air pollution exhibits both temporal and spatial variability, necessitating specialized preprocessing techniques to capture these dynamics accurately. This study presents a data preprocessing framework that applies temporal and spatial clustering to refine air quality monitoring data. Temporal clustering is used to group data according to time-based patterns, such as daily and seasonal fluctuations, helping to reveal periodic trends and anomalies. Spatial clustering, on the other hand, organizes data by geographic location, allowing for the identification of localized pollution patterns and sources. By combining these two clustering approaches, the preprocessing framework enables a comprehensive understanding of air pollution distribution, providing a foundation for subsequent modeling efforts. This approach is validated on a large-scale urban air quality dataset, showing improved data consistency and enhanced model performance in predicting pollutant levels.

5.1 Introduction

Air quality data are distinguished by notable characteristics in both time and space. With regard to temporality, the data demonstrate a high degree of contemporaneity and are susceptible to fluctuations on a seasonal and daily basis. For instance, pollutant concentrations are frequently elevated during winter and at the peak of morning and evening commuting hours. In terms of spatiality, air quality evinces pronounced localization across different regions, attributable to variations in geography and human activities. Consequently, pollution levels may exhibit considerable divergence between cities.

The application of data clustering methods in the field of air quality monitoring serves as an essential approach for in-depth analysis and interpretation of complex data sets. By organizing similar air quality data points into clusters, cluster analysis can identify pollution sources, relationships between monitoring stations, and

H. Liu et al., *Data Science in Air Quality Monitoring*, Engineering Applications of Computational Methods 23, https://doi.org/10.1007/978-981-96-5777-3_5

regional air quality characteristics. This approach not only reveals underlying patterns in the data but also helps to identify anomalies, thus enabling timely countermeasures to be taken. Common clustering algorithms, such as K-mean clustering and DBSCAN.

Common clustering analysis methods include partitional clustering and hierarchical clustering techniques (Hruschka and Covoes 2005; Covões et al. 2009). Partitional clustering methods divide the data into subsets and simultaneously construct all the clusters. Due to its simplicity and efficiency, this method is widely used in the K-means clustering algorithm. Additionally, hierarchical clustering uses a distance metric to evaluate the dissimilarity between data points in the clusters and follows an iterative approach. Ward's method is the most commonly used technique in hierarchical clustering. Jinming et al. (2015) estimated the linear temperature trend from 1962 to 2011 using surface temperature data from approximately 570 meteorological stations in China with K-means spatial clustering. The results of the study showed that the k-means technique is effective in discovering information hidden in the dataset and by using the clustering results as input, the predictive power can be improved. Geva (1999) established a new algorithm for hierarchical unsupervised fuzzy clustering for time series forecasting. In hierarchical clustering methods, Euclidean distance is used to measure the similarity between different objects. Kusiak and Li (2010) used a k-means cluster analysis algorithm to create five scenarios in the input space for short-term prediction of power generated by wind turbines at low wind speeds. Kim and Seo (2012) used fuzzy clustering in the SVR modeling process for wind power prediction.

An increasing number of clustering methods now combine deep neural networks and unsupervised clustering methods for feature downscaling and extracting features from input data. Richard et al. (2020) build a combination of convolutional autoencoders and k-medoids algorithms to perform time series clustering (Klampanos et al. 2018). Establishment of autoencoder-driven weather clustering for earthquake source estimation during nuclear events (Yang et al. 2024). Establishment of an autoencoder improved clustering method for wind speed prediction (Ryu et al. 2020). Establishment of Convolutional Autoencoder for Feature Extraction and Clustering for Smart Meter Load Analysis. Arasteh et al. (2024) used a hybrid approach combining autoencoder and K-means algorithm to develop a software fault predictor.

5.2 Data Acquisition

The Seoul area, as the capital of South Korea, is a significant industrial and commercial city. This dataset includes data from 25 stations located in districts such as North Jongno-gu, Jung-gu, Yongsan-gu, Eunpyeong-gu, Seodaemun-gu, Mapo-gu, Seongdong-gu, Gwangjin-gu, Dongdaemun-gu, and Gangdong-gu. Due to the dense population and heavy traffic in the Seoul area, air pollution is a major problem. According to air quality reports in recent years, Seoul has a high air pollution index among major cities, especially during winter and peak hours. Studying air quality monitoring in the Seoul region is important for public health because it can predict changes in the concentration of harmful pollutants in a timely manner and assess the impacts on the health of residents, especially the risk of respiratory and cardiovascular diseases. In addition, monitoring data can provide a basis for policy-makers to promote measures to reduce pollution sources, thereby improving overall air quality and safeguarding the health and well-being of the public.

To ensure the generalizability of this study, 8760 samples with a time interval of 1 h were used, covering the period from 2017-01-01 00:00 to 2017-12-31 24:00. This study analyzed six air quality variables, namely PM2.5, PM10, SO_2, NO_2, O_3, and CO. Since PM2.5, as one of the most typical air pollutants, this study predicts the future PM2.5 concentration values through the six previously mentioned variables. Each city has several monitoring stations, and the pollutant concentrations in the city are obtained by averaging the data from all the stations. The distribution of air quality testing stations in the Seoul area is shown in Fig. 5.1.

As can be seen in Fig. 5.1, all pollutant time series exhibit clear periodicity. In addition, based on the correlation between the variables, it can be seen from the heat map in Fig. 5.2 that PM2.5 and PM10 and SO_2 and NO_2 were found to be more similar.

These features indicate that the mean values of air quality data for Fig. 5.3.

STL decomposition is a method of decomposing a time series into three main components: trend, seasonality, and residuals. The method is particularly suitable for non-stationary time series with clear seasonality and trend, and series such as air pollutant concentrations are well suited for the application of STL decomposition. The results are given in Figs. 5.4 and 5.5.

PM2.5 and PM10 are representative air pollutants, and the random distribution of the residual components for these two variables indicates that most trends and

Fig. 5.1 Stations in the Seoul area

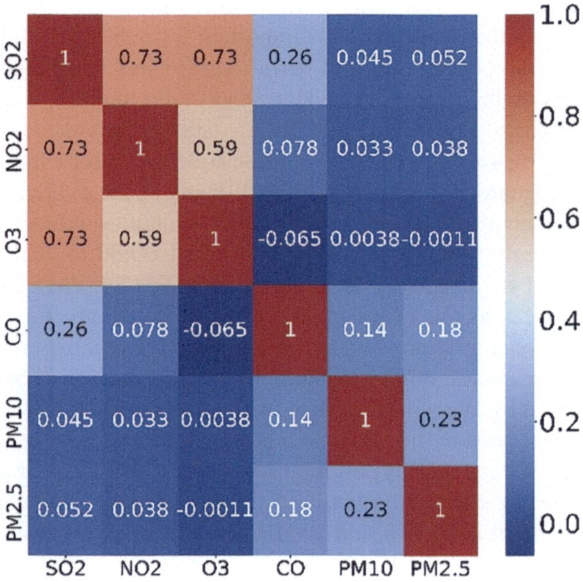

Fig. 5.2 Heat maps of pollutant variables

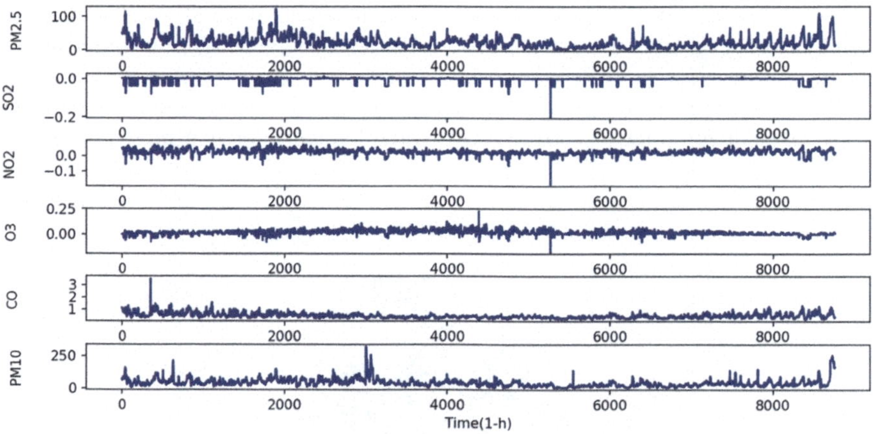

Fig. 5.3 Time series of variables related to pollutant concentrations

seasonal characteristics have been successfully captured. The trend component reveals the long-term trends in PM2.5 and PM10 concentrations, with significant increases in both during the spring and winter. The seasonal component displays a repeating fluctuation pattern in PM2.5 and PM10 concentrations over the cycle, reflecting seasonal influences, with higher concentrations in winter and spring, and lower concentrations in summer and fall.

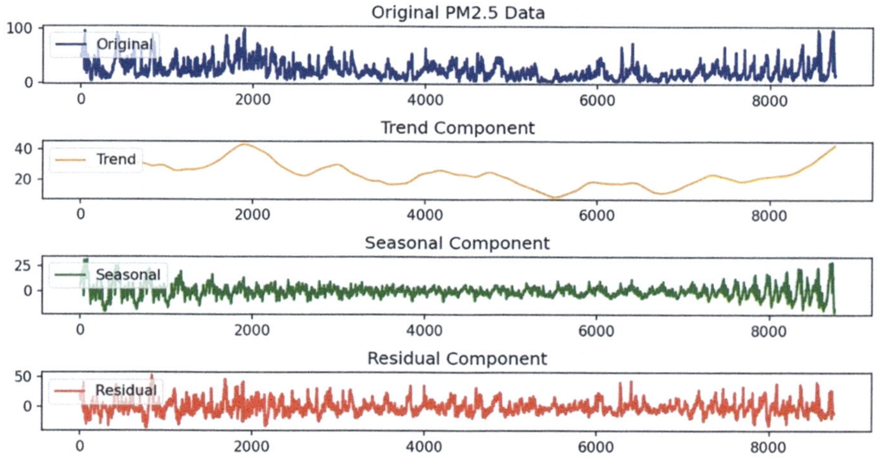

Fig. 5.4 STL decomposition of PM2.5 pollutant time series

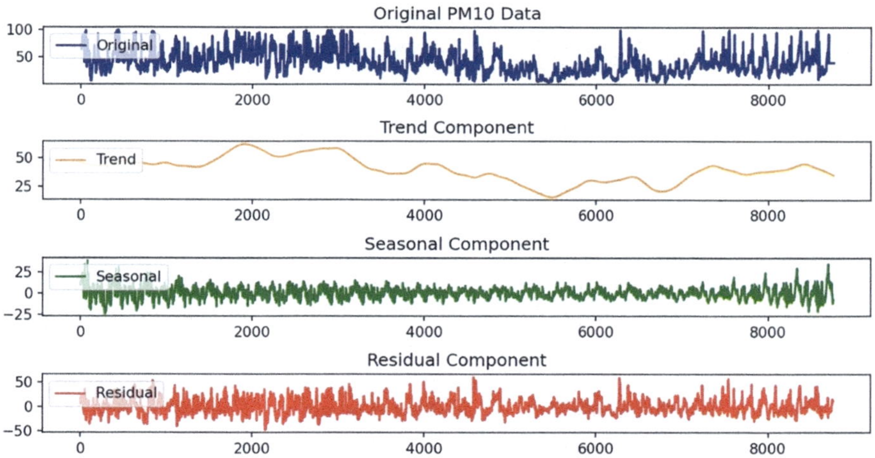

Fig. 5.5 STL decomposition of PM10 pollutant time series

5.3 Temporal Clustering of Air Quality Data

5.3.1 Definition and Role of Temporal Clustering

5.3.1.1 Definition

Temporal clustering is the process of grouping periods (e.g., hours, days, weeks, etc.) based on similarities in time series data. It identifies groups that exhibit similar characteristics over certain periods by analyzing trends and patterns in the data over time. This clustering is usually based on the similarity of certain characteristics, such as pollutant concentrations, meteorological conditions, etc.

5.3.1.2 Role

Temporal clustering provides active support for air pollutant prediction by improving data quality, discovering potential relationships and optimizing input features. It is capable of removing noisy data, revealing similarities between periods, and identifying which periods have a significant impact on pollutant concentrations, thus optimizing model features. In addition, the clustering results can be used to customize the model by developing specific prediction strategies for different clusters, improving the accuracy and generalization of the model, and adapting it to seasonal variations.

5.3.2 DBSCAN Temporal Clustering

5.3.2.1 Basic Theory

DBSCAN is density-based clustering method with noise, which is a very typical density clustering algorithm. The so-called density is the aggregation of points, and by adjusting the parameters, the points that meet the density conditions can be clustered into one class. Unlike distance-based clustering algorithms such as K-means, DBSCAN can be applied to both clustering convex sample sets and non-convex sample sets and is therefore very suitable for clustering in the time dimension.

DBSCAN is an unsupervised clustering algorithm based on density, which determines the clustering structure based on the density of the sample distribution (Sander et al. 1998). BSCAN measures the density of the point space based on the number of adjacent points within the neighborhood. The density reachability relation derives a set of samples with the highest density connectivity, which forms a category of the final clusters, referred to as "clusters." The primary advantage of DBSCAN over distance-based clustering algorithms is its ability to find clusters of any shape without prior knowledge of the number of clusters. DBSCAN has two important parameters: the radius of the neighborhood scan ε (eps) and the minimum number of points (MinPts) required to form a cluster.

5.3.2.2 DBSCAN Arithmetic Process

Defining the Parameters

The DBSCAN algorithm has two key parameters:

- ϵ(eps): defines the maximum distance (radius) at which a data point is considered a neighbor.
- minPts: the minimum number of neighbors around a point that are considered as core points.

Classification of Core, Boundary, and Noise Points

- Core point: a point is a core point if the number of points in its ϵ-neighborhood (including itself) is greater than or equal to minPts.
- Boundary point: a point is not a core point itself, but lies within the ϵ-neighborhood of some core point.
- Noise point: a point that is neither a core point nor a boundary point.

Definition of ϵ-neighborhood

For a point p in each dataset D, define its ϵ-neighborhood (Epsilon Neighborhood) as the set of points that satisfy the following conditions

$$N_{\epsilon}\left(p\right)=\left\{q\in D|dist\left(p,q\right)\leq\epsilon\right\} \tag{5.1}$$

where $dist(p, q)$ is the distance between points p and q. Euclidean distance or other metrics can be used.

Clustering Process

The DBSCAN clustering algorithm follows the following steps:

Step1: Randomly select an unvisited point p.
Step2: Check whether p is a core point, i.e., check whether the number of points in its ϵ-neighborhood is at least minPts.
Step3: If p is a core point, extend a new cluster with p as the starting point.
Step4: If p is not a core point and the number of points in its neighborhood is less than minPts, it is labeled as a noise point (which may subsequently become a boundary point).
Step5: Repeat the above steps for each point q that belongs to the neighborhood of a core point to extend the new cluster until there are no new points to add to the cluster.
Step6: Repeat the above steps until all points are visited.

Overview of the Formula

In DBSCAN, the definition of core points satisfies the following conditions:

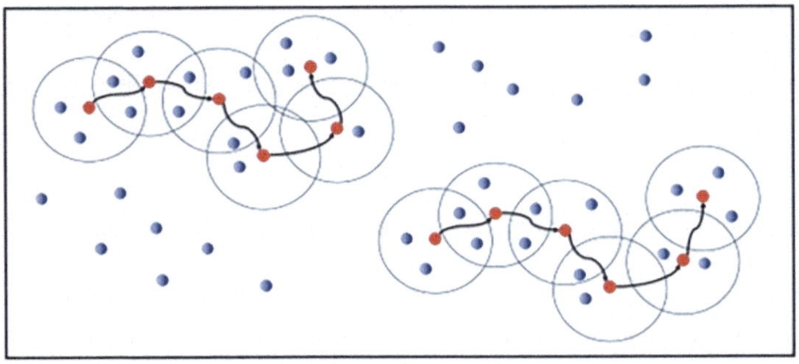

Fig. 5.6 Process of DBSCAN

$$|N_{\epsilon}(p)| \geq minPts \tag{5.2}$$

The step of clustering extension can be described as follows: if p is a core point, recursively add all unvisited points in its ϵ-neighborhood to the current cluster.

The clustering process of DBSCAN can be described as follows: a core point is randomly selected as the starting point, all density-reachable sample points belonging to that core point are identified, and a maximized region containing both core and boundary points is determined. When any two points within this region have the same density, a cluster is formed. Then, another core point is randomly selected, and the above process is repeated to form another cluster. These steps are repeated until all core objects are categorized. The above principle is visualized with the help of Fig. 5.6.

DBSCAN clustering results are given in Figs. 5.7 and 5.8:

5.3.3 *AE-DBSCAN Temporal Clustering*

5.3.3.1 Principle of AE

AE is a neural network for unsupervised learning tasks, especially for feature extraction and data dimensionality reduction. (Ang et al. 2016) Its core principle is to extract features and patterns of data by learning a low-dimensional representation (encoding) to reconstruct the input data as much as possible. The encoder's role is to map the high-dimensional input X into a low-dimensional hidden variable Z, thereby forcing the neural network to learn the most informative features. The decoder's role is to restore the hidden variables of the hidden layer to their original dimensions, and the ideal state is for the decoder's output to perfectly or approximately recover the original input., i.e., $X^R \approx X$, as shown in Fig. 5.9.

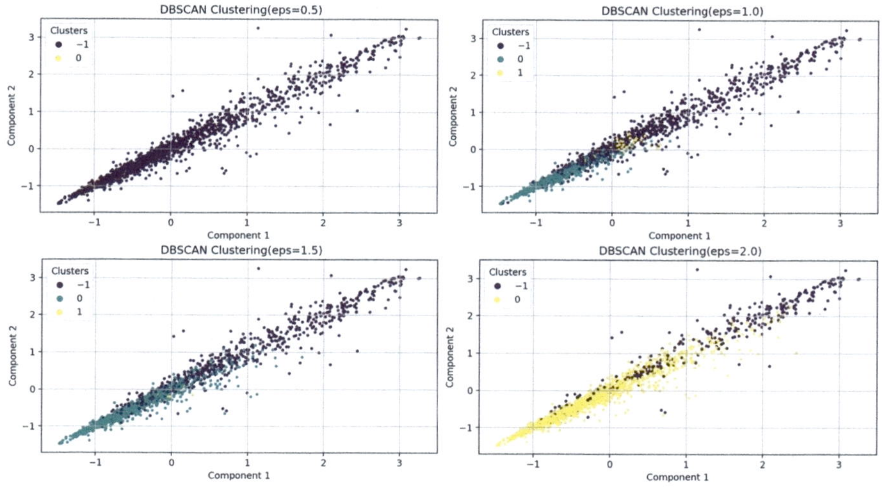

Fig. 5.7 DBSCAN clustering result

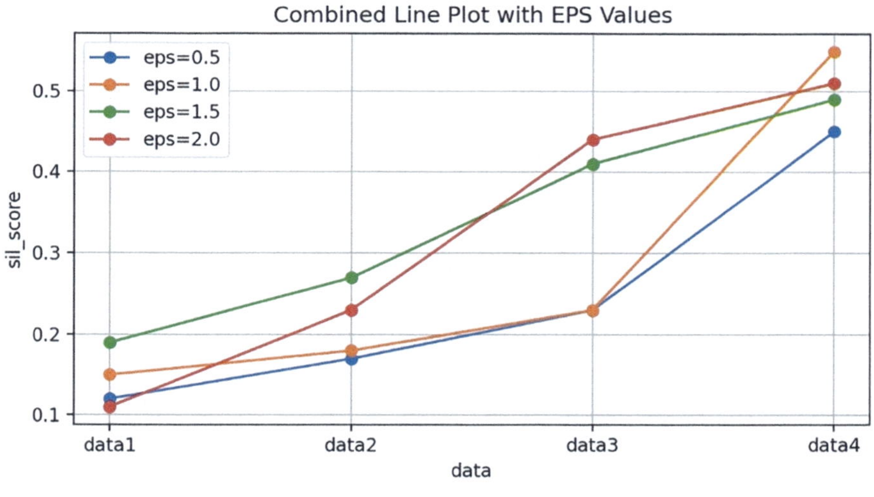

Fig. 5.8 Silhouette Score of different eps values

5.3.3.2 Modeling Step

Data Preparation

Collect air pollution time series data, which usually includes period (e.g., hours, days) characterization data, such as pollutant concentration, temperature, humidity, etc. Preprocess the data to normalize the air pollution time series data for model training.

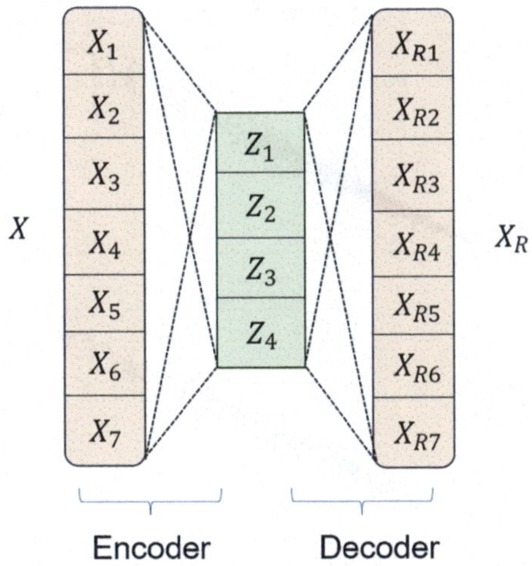

Fig. 5.9 AE structure figure

Constructing the Auto-Encoder (AE)

Step1: Construct the encoder and decoder parts of the autoencoder. The encoder maps the high-dimensional input data to the low-dimensional latent space layer by layer, while the decoder maps the low-dimensional representation back to the original data space.

Step2: Input time series data and train the self-encoder by minimizing the reconstruction error (e.g., mean square error) so that the model can reconstruct the input data.

$$L\left(x,\hat{x}\right)=\left|x-\hat{x}\right|^{2} \tag{5.3}$$

Step 3: Once the training is complete, the encoder is utilized to convert the time series data into a low-dimensional feature representation. The data for each period is compressed into a vector containing key features.

Step 4: Set the core parameters of DBSCAN and define the distance threshold between data points to determine whether the data points belong to the same cluster. Define the minimum number of data points MinPts required to form a dense region.

Step 5: Input the low-dimensional features output from the self-encoder into the DBSCAN algorithm. DBSCAN performs clustering based on the density of the data points, and is able to recognize clusters of arbitrary shapes, making it suitable for processing non-linearly distributed time series data.

Step 6: DBSCAN is also able to identify noisy points (outliers), which usually do not belong to any cluster and can be further analyzed or eliminated.

Step 7: Analyze the effect of DBSCAN clustering by visualizing the clustering results in a low-dimensional feature space. Differences between clusters can be demonstrated using 2D or 3D scatter plots.

Step 8: Analyze the period features in each cluster to understand the temporal patterns represented by the clusters. For example, differences between clusters for high pollution periods and low pollution periods can be found.

5.3.4 CAE-DBSCAN Temporal Clustering

5.3.4.1 Principle of CAE

Convolutional AutoEncoder is an unsupervised learning model based on convolutional neural networks (CNNs), which is mainly used for tasks such as feature extraction, dimensionality reduction, and denoising. (Sun et al. 2023) The core concept of CAE is to compress the input data into a low-dimensional representation using an encoder, and then reconstruct it using a decoder, with the aim of minimizing the reconstruction error between the inputs and outputs.

Encoder

- Role: compresses the input data into a low-dimensional feature representation (i.e., latent space representation), i.e., extracts the core features of the input data.
- Structure: The spatial dimensions and the number of feature maps are gradually reduced using convolutional and pooling layers, allowing the input data to be compressed into small-sized feature vectors. The convolutional layer is used to extract local features while the pooling layer is used for downsampling to reduce the dimensionality.
- Output: The final output of the encoder is a low dimensional feature vector, which is a simplified representation of the input data.

Decoder

- Role: Reduces the low dimensional feature representation generated by the encoder to the same dimensions as the input data, generating an output that is as similar as possible to the input.
- Structure: Gradually increase the spatial dimensions of the feature map using inverse convolutional (or upsampling) and convolutional layers to reconstruct the data with the same dimensions as the input.
- Output: the output of the decoder should match the shape of the input data, and the model makes the output as close as possible to the input data by minimizing the reconstruction error.

5.3.4.2 Modeling Step

Step1: Data preparation

Data collection: collect relevant datasets to ensure data completeness and accuracy.

Data Cleaning: Handle missing values and outliers to ensure data quality.

Data standardization: standardize or normalize the data as needed to eliminate scale differences between features.

Step2: Feature engineering

Feature selection: select the features that are meaningful for clustering.

Feature Extraction: Extract important features from raw data using feature extraction techniques (e.g., PCA or CAEs).

Data Enhancement: Enhance the data as needed to expand the dataset and improve the robustness of the model.

Step3: Construction of Convolutional Auto-Encoder (CAE)

Network architecture design: Select appropriate convolutional layer, activation function, pooling layer and fully connected layer to design the network architecture of CAE.

Step4: Training CAE:

Train the CAE using the collected data by minimizing the reconstruction error so that the network learns a valid representation of the data.

Step5: Feature Extraction:

Extract key features by converting the raw data into a low-dimensional representation through the trained CAE.

Step6: Application of DBSCAN clustering

Select the DBSCAN parameters:

Determine ε (neighborhood radius) and MinPts (minimum number of neighbors at core points), which can be selected by rule of thumb or parameter tuning.

Step7: Clustering Process:

Cluster the low-dimensional features of the CAE output using the DBSCAN algorithm to identify different clusters and noise points.

Step8: Evaluate the clustering results

Visualization: use dimensionality reduction techniques (e.g., t-SNE or UMAP) to visualize the clustering results, which facilitates intuitive understanding of the clustering effect.

Clustering quality assessment: Use appropriate assessment metrics (e.g., contour coefficient, Davies-Bouldin index) to assess the clustering effect.

Result analysis: Analyze the clustering results to understand the characteristics and meaning of each cluster and verify the effectiveness of the model.

Step9: Model Optimization

Parameter tuning: Adjust the parameters of CAE and DBSCAN according to the evaluation results to optimize the model performance.

Iterative improvement: Repeat the process of data preparation, feature engineering, model training and evaluation to gradually improve the clustering quality.

In the 1-step prediction using LSTM, the predicted and actual values are shown in Fig. 5.10. The results show that CAE-DBSCAN improves the prediction effect after clustering by combining the convolutional autoencoder and DBSCAN clustering. The self-encoder effectively extracts the key features of high-dimensional data, enhances the robustness of complex data distribution and noise, and reduces the risk of overfitting. In addition, the better feature representation allows subsequent prediction models to be trained on higher-quality data, which improves prediction accuracy. This combination of deep learning and traditional clustering fully utilizes the advantages of both, enabling CAE-DBSCAN to perform better in a variety of application scenarios.

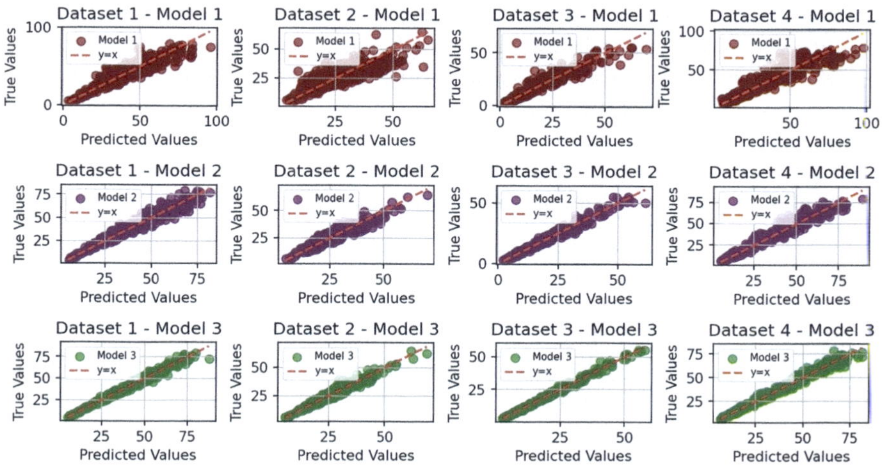

Fig. 5.10 1-step prediction using LSTM

5.4 Spatial Clustering of Air Quality Data

Spatial clustering is the process of categorizing data points in geographic space based on their similarity. It is commonly used to identify patterns and structures in the spatial distribution of data, particularly in fields such as Geographic Information Systems (GIS), environmental monitoring, and urban planning. Spatial clustering allows similar data points to be aggregated together based on a variety of characteristics (e.g., location, pollutant concentrations, meteorological data, etc.) in order to discover features and trends in spatially similar areas.

Spatial clustering provides active support for subsequent air pollutant predictions by identifying similarities and distribution patterns within geographic areas. It enhances data relevance and simplifies complexity, thereby improving model accuracy. In addition, spatial clustering helps to optimize feature selection by focusing the model on important regional features that affect prediction and supports the development of region-specific prediction models. By revealing spatial dependencies, cluster analysis better captures the interactions between regions.

5.4.1 K-Means Clustering

5.4.1.1 Basic Theory

The K-means algorithm is a clustering algorithm that is simple to implement and has been widely used in many fields (Pakhira 2009). As a partitional clustering algorithm, the key to classifying a given dataset into clusters is to find the least-square error between each data point in the dataset and the cluster mean, and then assign each data point to the cluster center closest to it.

Initially, the K-means algorithm randomly selects the specified k centers of mass from the dataset, evaluates the distance of each data point from all the selected centers of mass, and assigns each data point to the nearest center of mass as a member of the cluster for that center of mass. The centers of the cluster are re-evaluated when assigning new members to the cluster and the algorithmic process is executed in an iterative manner until the cluster membership is stable. The basic steps of the K-means algorithm for clustering contaminants in subway stations are as follows.

- Step1: Perform a random selection of k initial partitions on the station data.
- Step2: Generate a new clustering partition by assigning each station to the nearest cluster center.
- Step3: New clustering cluster center computation.
- Step4: Repeat Step2 and Step3 until the global average error is minimized and stabilized after site clustering.

From the above, it can be seen that the choice of the number of clusters k for the K-means clustering algorithm is crucial, and the performance of the algorithm depends on the specified value of k. Different values of k produce different results. In addition, the resulting clusters are also affected by the choice of the initial center of mass, different initial centers of mass produce different clusters, sometimes reaching a local minimum, so many iterations are needed to ensure better convergence. At the same time, the distance measure of clustering is also crucial, for the calculation of the distance from data points to the cluster center, the standard K-means algorithm uses the Euclidean distance measure.

In order to fully explore the spatial correlation between sites and ensure the optimal spatial features, the value of k is chosen to be 4, and five clusters are divided, while the number of iterations is set to 300. Station classification Centers are shown in Table 5.1, and the results are given in Figs. 5.11 and 5.12.

5.4.2 GMM

5.4.2.1 Principle of GMM

Gaussian Mixture Model (GMM) is a probabilistic model used to represent datasets with subpopulations of multiple Gaussian distributions (Wan et al. 2019). GMM assumes that the data points are generated from a mixture of multiple Gaussian

Table 5.1 Station classification center

	Result
C1	Station4, Station9
C2	Station1, Station2, Station3, Station6, Station7, Station11, Station15, Station17, Station18, Station19, Station21, Station22, Station23, Station24, Station25 Station8, Station10, Station12, Station13, Station14
C3	Station5, Station16
C4	Station20

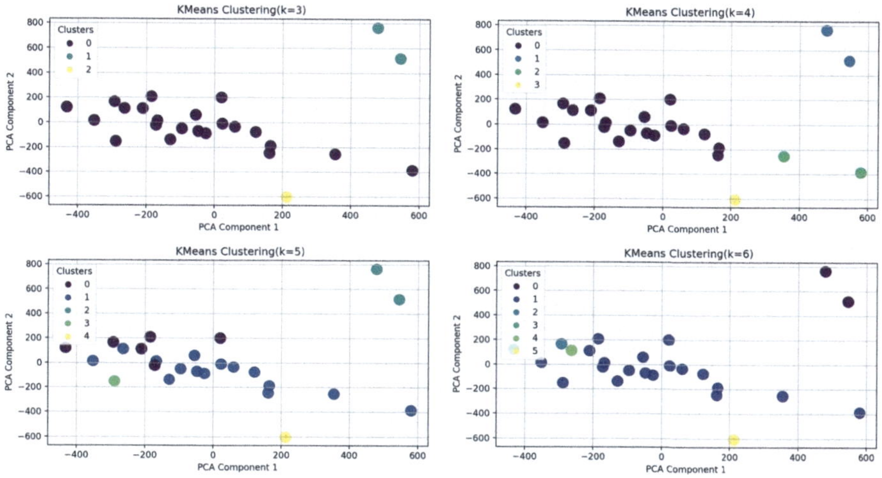

Fig. 5.11 Results of different k values

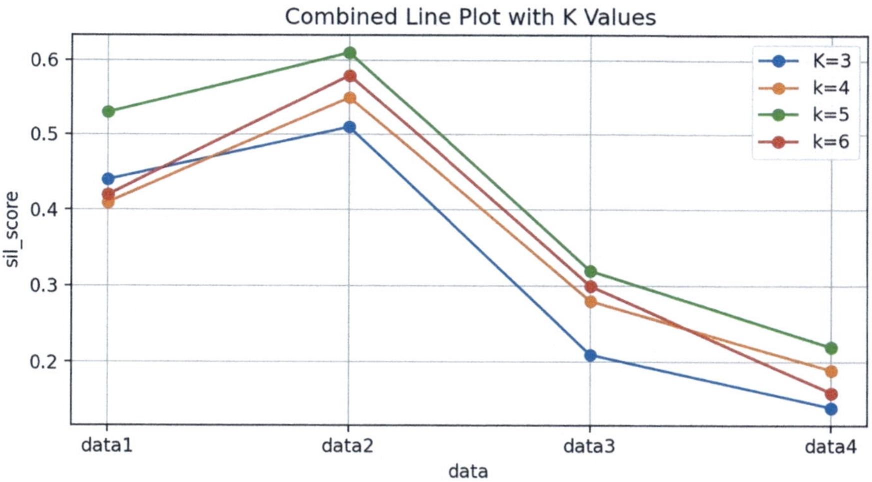

Fig. 5.12 Silhouette Score of different k values

distributions, with each Gaussian distribution corresponding to a cluster. Therefore, GMM is commonly used in clustering tasks for unsupervised learning and can be considered as a soft clustering of data as it assigns each data point a probability of belonging to each cluster. The following is the principle and main steps of GMM:

$$p(x) = \sum_{k=1}^{K} \pi_k \cdot N\left(x \mid \mu_k, \Sigma K\right) \tag{5.4}$$

5.4.2.2 Modeling Step

Step1: model assumptions

The GMM assumes that the dataset is generated from several different Gaussian distributions and defines each Gaussian distribution using the following parameters:

- Mean vector (μ): defines the center of the Gaussian distribution.
- Covariance matrix (Σ): defines the shape and size of the Gaussian distribution.
- Mixture coefficient (π): defines the weight of each Gaussian distribution, i.e. the probability that a data point belongs to that distribution.

Step2: Parameter Estimation: EM Algorithm

The GMM uses the Expectation Maximization (EM) algorithm to estimate the model parameters. The steps of the EM algorithm are as follows:

- Initialization: Randomly initialize the parameters of the model (mean, covariance matrix, mixing coefficients).
- Expectation step: Calculate the posterior probability that each data point belongs to each cluster (i.e., soft assignment probability) based on the current model parameters.
- Maximization: update the parameters of the model using the probabilities from the E step. The mean, covariance matrix, and mixing coefficients are re-estimated to maximize the log-likelihood of the data.

Repeat the above steps until convergence, i.e. no more significant changes in the parameters.

Step3: Cluster assignment

After several iterations of the EM algorithm, each data point will be assigned a cluster label based on its probability of belonging to different clusters. In GMM, each data point can have the probability of belonging to multiple clusters, not just to a single cluster (soft clustering).

5.4.3 GAE -Kmeans

5.4.3.1 Principle of GAE

Graph Autoencoder is an unsupervised learning method based on Graph Neural Networks (GNN) for learning node representations of graph structured data. GAE is mainly used for graph embedding learning, where the neighbor information or node features of the graph are reconfigured in the embedding space by means of a self-encoder structure in order to preserve the structural information and node attributes of the graph. The following is the working principle of graph self-encoder: Finally, the graph is fed into GCN to extract features in the spatial domain based on the spectral approach. It can be seen that there are two important components in the implementation of the GCN spatial feature extraction network, one is the construction of the graph and the other is the feature graph convolution operation.

Encoder

The encoder part encodes the input graph using a graph neural network such as Graph Convolutional Network (GCN). It extracts information from the node features and graph structure of the input graph and maps it into a low dimensional embedding space.

GCN can be computed in the encoder by the following equation:

$$H^{(l+1)} = \sigma\left(\tilde{D}^{-\frac{1}{2}}\tilde{A}\tilde{D}^{-\frac{1}{2}}H^{(l)}W^{(l)}\right) \tag{5.5}$$

where $\tilde{A} = A + I$ is the adjacency matrix with the addition of the self-loop, $\tilde{D}$ is the corresponding degree matrix, is the first l layer of the node feature matrix. $W^{(l)}$ is the trainable weight matrix. σ is the activation function.

Decoder

The task of the decoder is to reconstruct the neighbor matrix or similarity between nodes of the graph from the node embedding generated by the encoder. In GAE, the inner product decoder is usually used to estimate the relationship between two nodes by the following equation:

$$\hat{A}_{ij} = \sigma\left(Z_i^T Z_j\right) \tag{5.6}$$

where $\hat{A}_{ij}$ are the elements of the reconstructed adjacency matrix obtained by decoding, denoting the nodes. $Z_i^T Z_j$ is the node the inner product of the embedding vectors of node i and node j. σ is the activation function, and the Sigmoid function is usually chosen to ensure that the output values are between and suitable for interpretation as probability values.

GCN as shown in Fig. 5.13 The graphical convolution operation is used to extract spatial features, while the subsequently connected intelligent predictive model is used to extract temporal features.

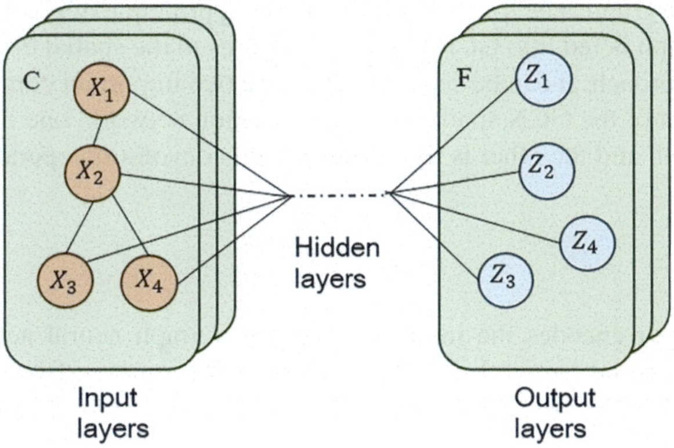

Fig. 5.13 GCN structure figure

5.4.4 *Modeling Step*

5.4.4.1 KMeans Clustering Modeling Process

Step1: Collect air quality or related data and organize them into a format suitable for cluster analysis, which usually includes multiple features (e.g., pollutant concentrations, meteorological conditions, etc.).

Step2: Identify features for clustering

Step3: Determine the appropriate number of clusters, K, by method (e.g., elbow method, profile coefficients, etc.).

Step4: Randomly select K initial clustering centers, assign data points to the closest cluster centers, and update the cluster centers to be the mean of the data points in their respective clusters. Repeat the above steps until the cluster centers do not change significantly or the maximum number of iterations is reached.

Step5: Analyze the features of each cluster to identify and visualize the contamination features in different areas.

5.4.4.2 GCN Extraction of Spatial Features Modeling Process

Step1: Prepare graph structure data, transform air quality data into a graph with nodes representing monitoring stations or areas and edges representing spatial relationships (e.g., distance or neighbors).

Step2: Define feature vectors for each node, e.g., pollutant concentration at monitoring stations, weather information, etc.

Step 3: Construct the adjacency matrix of the graph, which represents the connection relationship between nodes. K-nearest neighbor method or distance-based thresholding can be used to construct the adjacency matrix.

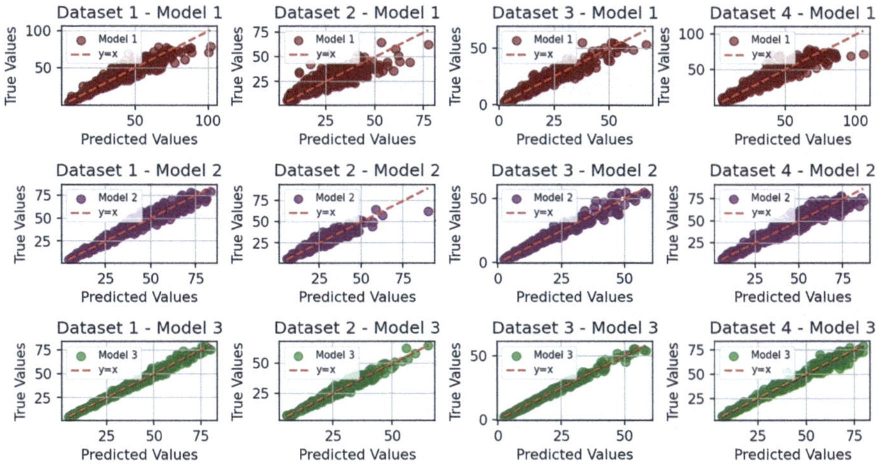

Fig. 5.14 1-step prediction using LSTM

Step4: Design the GCN model, including the input layer, hidden layer and output layer. The GCN uses the neighbor matrix and node features to perform convolution operation to extract spatial features.

Step5: In the forward propagation process, node features are propagated and aggregated through the adjacency matrix to update the feature representation of the nodes.

Step6: Calculate the loss function according to the task (e.g., regression or classification) and optimize the model parameters by back propagation.

Step7: Analyze the spatial features of the GCN output and combine the clustering results to further understand the influencing factors of regional air quality

The results of the experiment are shown in Fig. 5.14.

5.5 Clustering Performance Comparison

5.5.1 Evaluation with Silhouette Score

The Silhouette Coefficient is a measure of clustering quality that evaluates the similarity of each data point to the points within the cluster it belongs to, as well as the dissimilarity to the nearest cluster. The value of the silhouette coefficient ranges from −1 to 1. For a data point i, the contour coefficient $s(i)$ is defined as:

$$s(i) = \frac{b(i) - a(i)}{\max(a(i), b(i))} \tag{5.7}$$

Table 5.2 Silhouette score for different models

Silhouette score	Average result
DBSCAN	0.34
AE-DBSCAN	0.51
CAE-DBSCAN	0.63
K-means	0.39
GMM	0.41
GAE-Kmeans	0.59

$a(i)$ is the average distance from data point i to other points in the same cluster,$b(i)$ is the average distance from data point i to the nearest other cluster (Table 5.2).

5.5.2 Evaluation with Base Model

In this paper, LSTM is used as the base model to predict the clustering results for clustering performance comparison.

In this section, three commonly used statistical error assessment metrics—Mean Absolute Error (MAE), Root Mean Square Error (RMSE) and Mean Absolute Percentage Error (MAPE)—are used to quantitatively evaluate the prediction accuracy of the models. Table 5.3 summarizes these metrics.

The dataset in this chapter consists of 24-h data from 25 sites from January 1 to December 31, 2017, and predicts the average PM2.5 concentrations at the 25 sites. In order to evaluate the performance of the clustering model, the model uses four pollutant data for the spring, summer, fall, and winter seasons. The entire data set is 8760 (h). The entire dataset is 8760 h (2017.01.01 ~ 2017.12.31) of pollutant data, of which 0 ~ 2190 h (2017.01.01 ~ 2017.03.29) is dataset 1, 2190 ~ 4380 h (2017.03.29 ~ 2017.06.25) is dataset 2, and 4380 ~ 6570 h (2017.06.25 ~ 2017.09.22) as dataset 3, 6570 ~ 8760 h (2017.09.22 ~ 2017.12.31) as dataset 4. Each dataset is divided into training set and testing set according to 8:2.

This section carries out the comparison of the proposed temporal clustering, spatial clustering methods.

5.5.2.1 Comparison of Temporal Clustering

Model1 represents DBSCAN algorithm and Model2 represents AE-DBSCAB algorithm. Model3 represents CAE-DBSCAB algorithm.

After the calculation, the scatter plot of predicted and original data is shown in Fig. 5.15. From Fig. 5.16, it can be seen that the scattering points of Model2 are close to the diagonal line. This phenomenon indicates that the method can cluster spatial features more effectively.

CAE-DBSCAN performs better in two-step and three-step predictions due to its combination of convolutional autoencoder feature extraction capabilities, which

Table 5.3 Table of evaluation indicators

Evaluation indicators	Formula
MAE	$\frac{1}{N}\sum_{i=1}^{N}\lvert y_i - f_i \rvert$
RMSE	$\sqrt{\frac{1}{N}\sum_{i=1}^{N}(y_i - f_i)^2}$
MAPE	$\frac{100}{N}\frac{1}{N}\sum_{i=1}^{N}\left\lvert\frac{y_i - f_i}{y_i}\right\rvert$

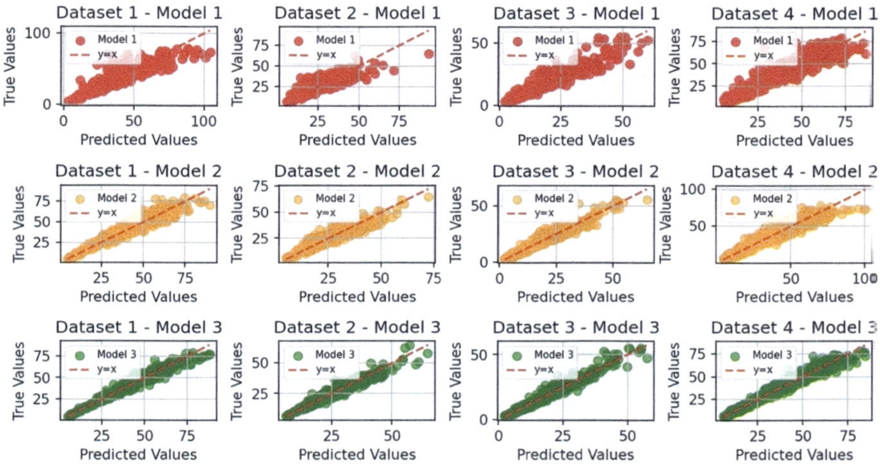

Fig. 5.15 2-step prediction using LSTM

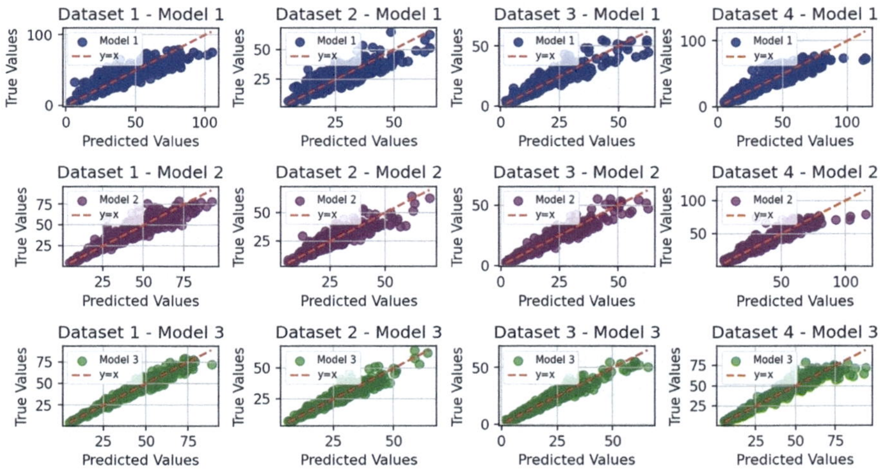

Fig. 5.16 3-step prediction using LSTM

allow it to capture deep features and improve clustering quality. At the same time, CAE-DBSCAN demonstrates greater robustness in handling noise and outliers, ensuring the stability of clustering results. Additionally, the high-quality clusters generated provide focused and reliable training samples for subsequent prediction models, enabling better capture of time series features and enhancing prediction accuracy. Its flexible adaptability allows the model to effectively adjust clustering structures in response to changes in data distribution, further improving prediction performance.

5.5.3 Comparison of Spatial Clustering

M1 represents the GAE -Kmeans algorithm and M2 represents the Kmeans algorithm, M3 represents the GMM algorithm, M4 represents the DBSCAN algorithm and M5 represents the AE-DBSCAN algorithm, M6 represents the CAE-DBSCAN algorithm. The results of the experiment are shown in Fig. 5.17.

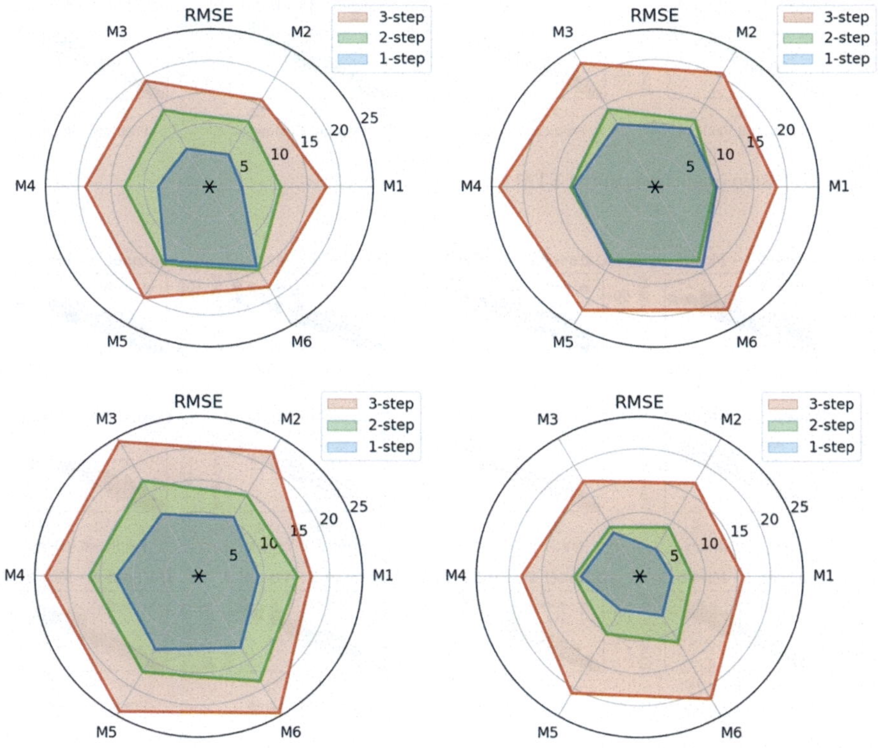

Fig. 5.17 Comparison of spatial clustering

GAE-KMeans performs better in two-step and three-step predictions due to its integration of graph autoencoder feature learning, which effectively captures the structural information of the data, enhancing feature representation and improving clustering quality. By combining graph convolutional networks with KMeans, GAE-KMeans can accurately identify the underlying clustering structures, resulting in more precise and consistent clustering outcomes that provide high-quality input data for subsequent prediction models. Additionally, its ability to handle complex relationships between data points ensures that similar data is clustered together, further optimizing the learning effectiveness of prediction models. With strong adaptability to varying data distributions and structures, GAE-KMeans delivers stable performance across diverse scenarios, ultimately enhancing the ability to capture time series features and improving prediction accuracy.

5.5.3.1 Comparison of Temporal-spatio Clustering

The method is to cluster the spatial clustering and then the temporal clustering and finally let the LSTM model be used for the prediction of PM2.5 concentration values will be Spatio-Temporal Clustering, Spatio clustering, Temporal Clustering three clustering, methods are compared in the following Table 5.4.

From the above table, it can be seen that spatial clustering alone can identify the distribution pattern of air pollutants in different geographic locations, but cannot capture the dynamic characteristics over time. Temporal clustering alone can reveal trends in air pollutants over time series, but lacks an understanding of geographic distribution. By considering both spatial and temporal features, a more comprehensive understanding of pollutant behavioral patterns, such as pollution trends in a given area over a specific period, can be achieved.

Combining spatial and temporal clustering allows complex patterns and relationships to be recognized. For example, changes in pollutant concentrations in different areas over a specific period may be influenced by common factors, such as meteorological conditions, traffic patterns, and so on. This can more accurately capture the combined effects of these factors on pollutant concentrations.

By clustering in both spatial and temporal dimensions, richer feature sets can be generated for model training. For example, time-series data for a specific area can be combined with historical clustering information for that area, allowing the model to better capture past pollutant behavior and thus improve its ability to predict future pollutant concentrations.

Air pollutant concentrations are affected by a variety of factors, including anthropogenic activities and natural factors. Combining spatial and temporal clustering can better capture these dynamics.

Table 5.4 Comparison of evaluation indicators

Prediction steps	Dataset	Model	RMSE	MAE	MAPE
Step1	Data1	**Spatio-temporal clustering**	**6.44**	**5.66**	**0.15**
		Spatio clustering	9.38	6.49	0.20
		Temporal clustering	10.53	8.23	0.24
	Data2	**Spatio-temporal clustering**	**3.66**	**2.28**	**0.09**
		Spatio clustering	4.95	3.55	0.15
		Temporal clustering	5.90	4.66	0.19
	Data3	**Spatio-temporal clustering**	**3.91**	**2.34**	**0.23**
		Spatio clustering	4.92	3.67	0.42
		Temporal clustering	4.93	3.68	0.39
	Datat4	**Time-space clustering**	**6.56**	**4.99**	**0.10**
		Spatio clustering	9.13	5.71	0.18
		Temporal clustering	10.68	8.05	0.62
Step2	Data1	**Spatio-temporal clustering**	**8.63**	**7.66**	**0.25**
		Spatio clustering	9.00	8.77	0.30
		Temporal clustering	12.13	9.78	0.34
	Data2	**Spatio-temporal clustering**	**9.11**	**8.34**	**0.29**
		Spatio clustering	10.90	9.18	0.31
		Temporal clustering	11.94	9.58	0.41
	Data3	**Spatio-temporal clustering**	**6.87**	**5.33**	**0.27**
		Spatio clustering	7.98	6.63	0.42
		Temporal clustering	8.93	3.68	0.39
	Datat4	**Time-space clustering**	**7.89**	**6.55**	**0.21**
		Spatio clustering	15.13	13.71	0.56
		Temporal clustering	14.68	11.05	0.72
Step3	Data1	**Spatio-temporal clustering**	**10.44**	**9.66**	**0.45**
		Spatio clustering	18.42	16.69	0.67
		Temporal clustering	20.53	19.57	0.79
	Data2	**Spatio-temporal clustering**	**11.64**	**10.14**	**0.35**
		Spatio clustering	17.85	16.59	0.59
		Temporal clustering	15.91	14.46	0.44
	Data3	**Spatio-temporal clustering**	**13.98**	**12.14**	**0.43**
		Spatio clustering	15.81	3.67	0.42
		Temporal clustering	16. 88	3.68	0.39
	Datat4	**Time-space clustering**	**16.56**	**14.49**	**0.56**
		Spatio clustering	17.16	15.81	0.67
		Temporal clustering	22.44	20.05	0.87

5.6 Conclusions

This section discusses methods of spatial and temporal clustering applied to air pollutant prediction, focusing on pollutant concentration prediction in Seoul subway stations. The dataset is divided into four groups for experimental validation, and the results show that combining these methods leads to higher prediction accuracy. The main findings are as follows:

- the spatial clustering method effectively extracts spatial features of air pollutant time series, and incorporating an autoencoder module and graph attention layer before the model enhances prediction accuracy.
- the temporal clustering method effectively captures temporal features of pollutant time series, and adding an autoencoder module and convolutional layer improves model accuracy.
- combining temporal and spatial clustering significantly enhances prediction accuracy and model robustness.

References

Ang JC, Mirzal A, Haron H, Hamed HNA (2016) Supervised, unsupervised, and semi-supervised feature selection: a review on gene selection. IEEE/ACM Trans Comput Biol Bioinform 13(5):971–989

Arasteh B, Golshan S, Shami S, Kiani F (2024) Sahand: a software fault-prediction method using autoencoder neural network and K-means algorithm. J Electron Test 40(2):229–243

Covões TF, Hruschka ER, de Castro LN, Santos ÁM (2009) A cluster-based feature selection approach. DBLP

Geva AB (1999) Hierarchical-fuzzy clustering of temporal-patterns and its application for time-series prediction. Pattern Recogn Lett 20(14):1519–1532

Hruschka ER, Covoes TF (2005) Feature selection for cluster analysis: an approach based on the simplified Silhouette criterion. In: International conference on computational intelligence for modelling, control & automation, & international conference on intelligent agents, web technologies & internet commerce

Jinming F, Yonghe L, Zhongwei Y (2015) Analysis of surface air temperature warming rate of China in the last 50 years (1962–2011) using k-means clustering. Theor Appl Climatol 120:785–796

Kim SJ, Seo IY (2012) A clustering approach to wind power prediction based on support vector regression. Int J Fuzzy Logic Intell Syst 12(2):108–112

Klampanos IA, Davvetas A, Andronopoulos S, Pappas C, Ikonomopoulos A, Karkaletsis V (2018) Autoencoder-driven weather clustering for source estimation during nuclear events. Environ Model Softw 102:84–93

Kusiak A, Li W (2010) Short-term prediction of wind power with a clustering approach. Renew Energy 35(10):2362–2369

Pakhira MK (2009) A modified k-means algorithm to avoid empty clusters. International Journal of Recent Trends in Engineering 1(1):220–226

Richard G, Grossin B, Germaine G, Hébrail G, De Moliner A (2020) Autoencoder-based time series clustering with energy applications

Ryu S, Choi H, Lee H, Kim H (2020) Convolutional autoencoder based feature extraction and clustering for customer load analysis. IEEE Trans Power Syst 35(2):1048–1060

Sander J, Ester M, Kriegel HP, Xu X (1998) Density-based clustering in spatial databases: the algorithm GDBSCAN and its applications. Data Mining Know Discov 2(2):169–194

Sun D, Liu L, Luo B, Ding Z (2023) GLASS: a graph Laplacian autoencoder with subspace clustering regularization for graph clustering. Cogn Comput 15(3):803–821

Wan H, Wang H, Scotney B, Liu J (2019) A novel gaussian mixture model for classification. In: 2019 IEEE international conference on systems, man and cybernetics (SMC)

Yang M, Guo Y, Fan F, Huang T (2024) Two-stage correction prediction of wind power based on numerical weather prediction wind speed superposition correction and improved clustering. Energy 302:131797

Chapter 6
Data Forecasting in Air Quality Monitoring

Abstract Accurate forecasting of PM2.5 concentrations plays a pivotal role in mitigating health risks and optimizing air quality management strategies, especially in regions with severe pollution challenges. This study evaluates the performance of five deterministic and two probabilistic forecasting models, leveraging diverse datasets from Changsha and Seoul to account for varying urban air quality dynamics. BiLSTM and Transformer consistently performed the best of the deterministic models, achieving the lowest MAE, RMSE, and MAPE. In contrast, ELM showed the highest error rates, indicating its limitations in capturing the complexities of air quality data. In the probabilistic forecasting domain, QRNN outperformed BNN in Changsha by providing more accurate prediction intervals, while BNN demonstrated superior reliability in Seoul despite wider intervals. These results highlight the importance of model selection based on dataset characteristics and environmental context, emphasizing the need for both deterministic and probabilistic approaches to enhance the accuracy and adaptability of air quality forecasting. Ultimately, these improvements will support better decision-making in air quality management.

6.1 Introduction

Air pollution has emerged as a critical global environmental issue, directly affecting human health, ecosystems, and climate change. Its sources are diverse, encompassing industrial emissions, vehicular transport, construction activities, and the use of biomass fuels (Wu et al. 2023). With accelerated urbanization, the growing number of vehicles and industrial activities has exacerbated outdoor air quality issues (Elbaz et al. 2023). However, indoor air pollution is equally significant, as individuals now spend the majority of their time indoors. In some cases, indoor pollution may pose even greater health risks. Key contributors include poor ventilation, indoor fuel combustion, volatile organic compounds (VOCs) from cleaning products, and harmful gases released from building materials (Kim et al. 2012). Prolonged exposure to polluted indoor environments has been associated with allergic reactions,

H. Liu et al., *Data Science in Air Quality Monitoring*, Engineering Applications of Computational Methods 23, https://doi.org/10.1007/978-981-96-5777-3_6

chronic respiratory diseases, and an increased risk of cardiovascular conditions (Liu et al. 2018). Despite the establishment of indoor air quality standards in many countries aimed at controlling pollutant levels, the issue remains inadequately addressed in practice (Son et al. 2014). As a result, comprehensive monitoring and forecasting of air quality, encompassing both indoor and outdoor environments, has become imperative.

In recent years, numerous models have been developed to forecast concentrations of key pollutants such as AQI (Liu et al. 2020), $PM_{2.5}$, PM_{10}, CO, NO_2, SO_2, and O_3. These models are generally classified into three categories: numerical models, statistical models, and intelligent models. Numerical models simulate pollutant diffusion and transmission processes through computational methods. While numerical models can effectively capture abrupt weather changes and offer broad predictive coverage, they often suffer from computational delays and lower accuracy. Statistical models, which rely on historical observational data, have been more effective at uncovering hidden trends in pollution series. Widely used models include the Autoregressive Integrated Moving Average (ARIMA) and machine learning models (Darekar and Reddy 2017; Fang et al. 2023). The latter, particularly AI-based models, have gained popularity in recent years due to their superior performance. A growing number of hybrid models have also been developed (Duan et al. 2018). Studies have consistently demonstrated the advantages of hybrid models in improving the generalization and precision of air quality forecasts.

Despite these advances, several research gaps remain. While machine learning and hybrid models have proven effective, they often require substantial computational resources and may face limitations in real-time applications. Additionally, the majority of models focus on outdoor air quality, leaving a gap in predictive accuracy for indoor environments where pollutant sources and dynamics are distinct (Su et al. 2023). There is also a need for improved methods to account for environmental variability and uncertainty in pollution forecasts, particularly during sudden pollution events (Tan et al. 2022).

Recent advancements in deep learning have shown promise in addressing these challenges. Extreme Learning Machine (ELM), with its rapid training capability, is well-suited for real-time monitoring. Gated Recurrent Units (GRU) and Bidirectional Long Short-Term Memory Networks (BiLSTM), which are adept at capturing temporal dependencies, offer robust solutions for dynamic air quality forecasting. Additionally, deep models such as the Deep Extreme Learning Machine (DELM) (Tissera and McDonnell 2016) and Transformer networks (Chen et al. 2023) leverage multilayer architectures and self-attention mechanisms to enhance predictive performance across large datasets.

To address the inherent unpredictability of pollution events, relying solely on deterministic forecasting methods proves to be insufficient (Nowotarski and Weron 2018). While these traditional models provide single-point predictions, they fail to account for the uncertainties associated with volatile and complex environmental phenomena (Bazionis and Georgilakis 2021). Such limitations highlight the importance of integrating probabilistic forecasting models into air quality monitoring systems. Unlike deterministic models, probabilistic models like Bayesian Neural

Networks (BNN), DeepAR, and Quantile Recurrent Neural Networks (QRNN) go beyond simple point estimates by offering a comprehensive range of possible outcomes through the generation of predictive distributions (Zhang et al. 2014). The deterministic and probabilistic prediction literature reviews are shown in Tables 6.1 and 6.2, respectively.

Table 6.1 A summary of the reviewed deterministic forecasting literature

Model	Contribution	Disadvantage
Mi et al. (2017)'s model	A combination of wavelet, ARMA, and ELM was proposed to improve forecasting accuracy by utilizing wavelet decomposition and outlier correction for more robust predictions	Despite this, the model's performance has been shown to decline in highly noisy or fluctuating datasets, which may affect its overall robustness
Yin et al. (2021)'s model	The hybrid forecasting method of ELM was proposed to combine boosting algorithms and ECM, which has been shown to improve both multi-step and hourly $PM_{2.5}$ predictions	Its prediction accuracy was observed to decrease as the forecasting steps increased, and more complex models may be needed
Fang et al. (2023)'s model	Fang's model was developed to reduce training parameters and accelerate convergence, which has contributed to better handling of multichannel input for air quality predictions	Its slower convergence and increased complexity in different environments have posed challenges for real-time applications
Guo et al. (2023)'s model	RF, CNN, and GRU were integrated to handle incomplete data, which has demonstrated superior time-series prediction accuracy over traditional models in $PM_{2.5}$ prediction tasks	Its generalization ability has been limited by small sample sizes, and its performance may degrade without additional measured variables
Ma et al. (2021)'s model	GCN and BiLSTM were combined to simultaneously capture spatial and temporal dependencies, leading to improved accuracy for predicting subway passenger flow compared to single models	Despite its enhanced accuracy, its complex model structure has been associated with higher computational costs and difficulties in hyperparameter tuning
Jia et al. (2022)'s model	The improved sparrow search algorithm was applied to optimize DELM, which has significantly enhanced the prediction accuracy and robustness	Its performance has been found to be less effective under non-random load conditions and could be sensitive to domain-specific data
Li et al. (2022)'s model	The IE-SBiGRU model was proposed to enhance long-term dependency handling and computational efficiency by reducing the attention matrix sparsity, while strengthening local correlations with BiGRU	Its generalization to other datasets was found to be limited, and its computational complexity increased with the inclusion of recurrent layers
Ren et al. (2023)'s model	The hybrid model was developed by integrating Informer and Encoder Forest to improve prediction accuracy and mitigate noise effects using decomposition techniques	Despite its improvements in accuracy, the hybrid approach was observed to be sensitive to the quality of data decomposition and may overfit when noise is poorly handled

Table 6.2 A summary of the reviewed probabilistic forecasting literature

Model	Contribution	Disadvantage
Wang et al. (2024)'s model	Wang's model was proposed to capture both fuzzy and probabilistic relationships by utilizing Bayesian networks for structure learning, improving its ability to model complex time series data	Its performance was found to be dependent on the quality of the Bayesian network structure and may suffer from increased computational cost
Luo et al. (2024)'s model	Luo's model was designed to enhance short-term probabilistic forecasting by leveraging accumulated hidden layer connections, improving forecasting accuracy, and handling data anomalies effectively	The model can still struggle with overfitting when applied to highly volatile datasets, and its computational complexity remains a challenge
Fang and Liu (2023)'s model	Fang's model was introduced to predict PM2.5 concentrations in subway stations by incorporating multi-resolution attention mechanisms, enhancing adaptability to complex environmental changes	Despite its improvements in handling uncertainty, the model was found to have difficulty managing large data outliers and error accumulation in long-term predictions
Li et al. (2023)'s model	Li's model was developed to improve mid-term probabilistic forecasting by incorporating quantile constraints and replacing LSTM with GRU to better capture long-term dependencies	Its complexity can limit real-time applications, and the accuracy of prediction intervals tends to decline significantly when the forecasting horizon is extended

These models not only predict the most likely values but also quantify the uncertainty surrounding those predictions by providing confidence intervals or probability distributions. This feature is particularly valuable in situations where pollution levels can fluctuate drastically due to unforeseen events such as extreme weather conditions, changes in traffic patterns, or industrial activities. By incorporating uncertainty into the forecasting process, probabilistic models improve the accuracy and dependability of air quality predictions, enabling decision-makers to evaluate risks more effectively and implement informed interventions.

6.2 Data Acquisition

This study utilizes experimental time series data from air pollution monitoring in two distinct regions to compare the performance of different forecasting models. Both datasets focus on $PM_{2.5}$, a widely used indicator in air pollution research due to its standardized, comprehensive, and timely data collection. The data resolution for the experiment is aggregated to daily measurements, and the visual trends and detailed characteristics are presented in Figs. 6.1 and 6.2; Table 6.3.

Specifically, the dataset for each region contains 2000 data points, which are divided into training and testing sets. The first 1600 data points constitute the training set, while the remaining 400 data points are used for testing. The Changsha ground-level pollution data spans from October 2, 2021, to December 27, 2021,

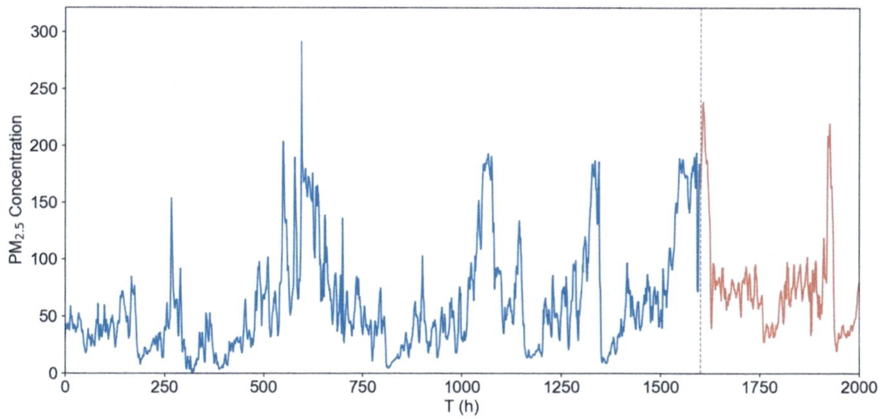

Fig. 6.1 The curve plots of the air quality data in Changsha

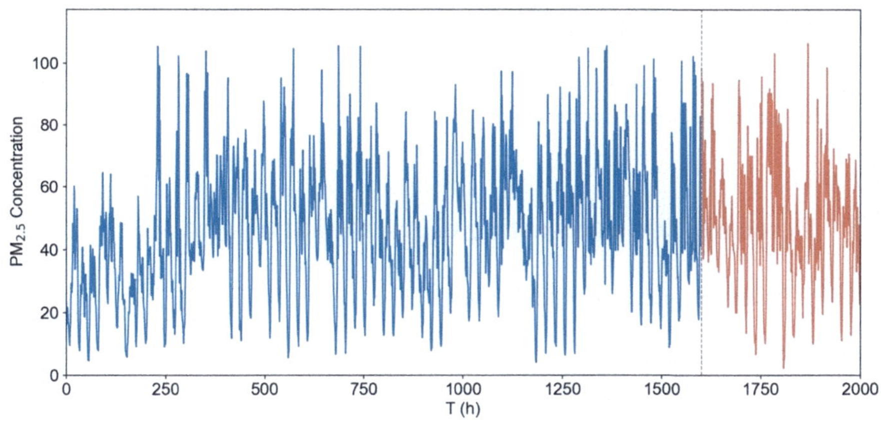

Fig. 6.2 The curve plots of the air quality data in Seoul

Table 6.3 The statistical descriptions of four $PM_{2.5}$ data sets

Data set	City	Min	Max	Mean	Standard deviation	Skewness	Kurtosis	Sample entropy
#1	Changsha	1	292	64.13	45.91	1.38	1.63	0.39
#2	Seoul	2.4	106	46.69	21.08	0.27	−0.33	1.24

while the indoor pollution data from Korea's Samsung subway station covers the period from January 22, 2023, to April 15, 2023. Both datasets are recorded at hourly intervals.

From the visual analysis and statistical characteristics, several patterns emerge. The Changsha dataset exhibits a higher mean pollution level with greater variability, a right-skewed distribution, and elevated kurtosis. These features suggest frequent occurrences of extreme pollution events. In contrast, the Korean subway dataset shows a lower average pollution level, reduced variability, and a more symmetrical distribution. The Korean data appears more stable, with fewer extreme values, likely due to stricter environmental controls or the relatively consistent nature of indoor environments.

In terms of pollution levels and variability, the Changsha dataset reveals higher pollution levels with significant fluctuations, potentially linked to industrial activities, heavy traffic, and ongoing urbanization. On the other hand, the Korean dataset reflects a more stable pollution pattern with lower levels of fluctuation, possibly due to controlled underground conditions and the absence of significant external pollution sources. Regarding the distributional characteristics, the Changsha dataset is right-skewed with high kurtosis, indicating frequent extreme values, while the Korean dataset demonstrates a flatter, more symmetric distribution, with fewer extreme pollution events. Finally, in terms of time series complexity and predictability, the Changsha dataset is characterized by low sample entropy and high autocorrelation. Conversely, the Korean dataset shows higher sample entropy, indicating a more complex structure despite its moderate autocorrelation, which may introduce challenges in predictive modeling.

6.3 Deterministic Forecasting of Air Quality Data

6.3.1 Extreme Learning Machine

6.3.1.1 Theoretical Basis

The ELM, a type of single-layer feedforward neural network, is primarily used for classification and regression tasks. Unlike traditional neural networks that adjust weights iteratively using methods like gradient descent, ELM randomly initializes the input weights and biases of hidden neurons, keeping them fixed during training. Only the output weights are determined analytically. Its core concept is that the hidden layer can transform input data into a sufficiently large and diverse feature space to approximate the target output without requiring parameter optimization for the hidden layer. The structure of ELM is depicted in Fig. 6.3.

One of the theoretical advantages of ELM is its universal approximation capability, meaning it can approximate any continuous function with sufficient hidden neurons. The output weights are typically solved using a closed-form solution, such as least squares, which significantly speeds up the training process. This makes ELM particularly effective for tasks that require rapid training and testing, especially with large datasets.

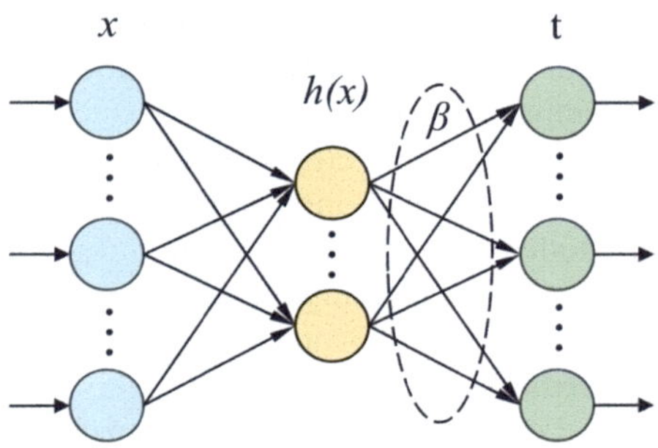

Fig. 6.3 The structure of the ELM

Assuming the input vector for the ELM is denoted as $\{x_i | i = 1, 2, \ldots, n\}$ and the output vector is $\{T_j | j = 1, 2, \ldots, m\}$, the network's output can be formulated as:

$$T = \left[t_1, t_2, \ldots, t_j \right]_{n \times m} = \left[\sum_{i=1, j=1}^{l, m} \beta_{ij} g\left(w_{ij} x_i + b_i \right) \right]_{n \times m} = \left(H_{n \times 1} \beta_{1 \times m} \right) \tag{6.1}$$

where $g(x)$ represents the activation function in the hidden layer neurons. The terms w_{ij} and b_i stand for the weights and biases between the input and hidden layers, respectively. The parameter l refers to the number of hidden neurons, while H is the hidden layer's output matrix, and β_{ij} signifies the weights connecting the hidden and output layers.

6.3.1.2 Key Features

Random Initialization with Analytical Solutions: ELM is characterized by its random initialization of hidden layer weights and biases, removing the need for iterative adjustments. The output weights are determined through an analytical solution, making the training process extremely fast compared to traditional neural networks.

Universal Approximation Capability: Despite its simplicity, ELM maintains a high degree of approximation power due to its capacity to map input data to higher-dimensional feature spaces. This allows ELM to effectively model complex nonlinear relationships, even with a single hidden layer.

6.3.1.3 Modeling Step

1. Random Initialization:
   The weights and biases for the hidden neurons are randomly assigned without iterative adjustments.
2. Input Transformation:

Input data is mapped to a higher-dimensional feature space through the activation function applied to the hidden layer.

3. Output Weight Calculation:

   The output weights are computed in a single step using a closed-form solution, often involving the Moore-Penrose inverse.

4. Model Training Completion:

   After calculating the output weights, the training process is complete, requiring no further updates or iterations.

5. Prediction:

   For new data, the input is transformed via the hidden layer and mapped to the output using the learned weights.

6.3.2 *Gated Recurrent Unit*

6.3.2.1 Theoretical Basis

The GRU, a simplified variant of the Long Short-Term Memory (LSTM) network, reduces computational complexity by using two gates and eliminating the separate cell state. Figure 6.4 illustrates the structure of the GRU network model employed in this chapter. The update gate combines the functions of the LSTM's forget and input gates, determining how much information from the previous hidden state is preserved. The reset gate, which uses different weight matrices, controls how much of the previous hidden state is forgotten. By merging the cell state with the hidden state and introducing other modifications, the GRU reduces parameters and computational costs. The primary formula for GRU is as follows:

$$z_t = \sigma\left(W_z^{(h)} h_{t-1} + W_z^{(x)} x_t + b_z\right) \tag{6.2}$$

$$r_t = \sigma\left(W_r^{(h)} h_{t-1} + W_r^{(x)} x_t + b_r\right) \tag{6.3}$$

$$\tilde{h}_t = \tanh\left(W_h^{(m)}\left(r_t \odot h_{t-1}\right) + W_h^{(x)} x_t + b_h\right) \tag{6.4}$$

$$h_t = \left(1 - z_t\right) \odot h_{t-1} + z_t \odot \tilde{h} \tag{6.5}$$

At time step t, x_t is the input vector, h_{t-1} is the hidden state vector from the previous step, z_t is the update gate vector determining the proportion of the previous hidden state h_{t-1} is retained, and r_t is the reset gate vector controlling the portion of h_{t-1} to be forgotten.

The candidate hidden state vector represents the potential new hidden state based on the current input and the reset hidden state. $W_z^{(h)}, W_z^{(x)}, W_r^{(h)}, W_r^{(k)}, W_h^{(m)}, W_h^{(x)}$ are the weight matrices for the update and reset gates and the candidate hidden state, applied to the hidden state and input, b_z, b_r, b_h are the bias vectors for the update gate,

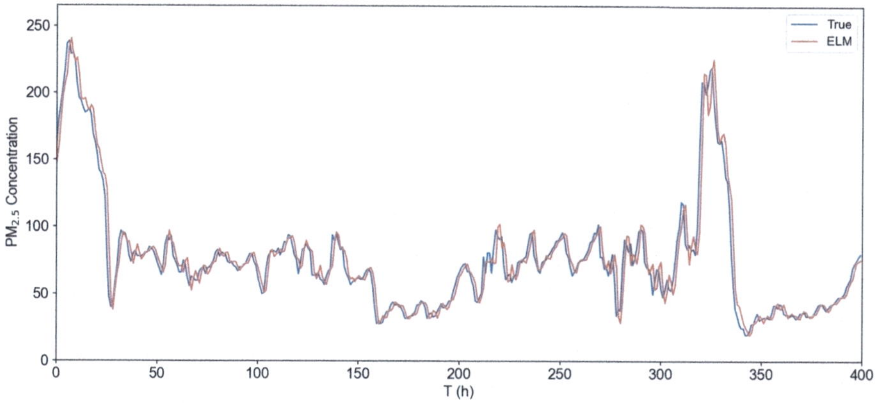

Fig. 6.4 The structure of the GRU

reset gate, and candidate hidden state, respectively. σ denotes the Sigmoid activation function, tanh is the hyperbolic tangent activation function, and $\odot$ represents element-wise multiplication.

6.3.2.2 Key Features

Fewer Parameters for Efficient Learning: GRU reduces the complexity of recurrent neural networks by merging the forget and input gates into a single update gate. This results in fewer parameters and faster training times compared to more complex architectures like LSTM, while maintaining the ability to capture sequential dependencies.

Flexible Control Over Sequence Memory: GRU introduces a gating mechanism that dynamically controls the retention and update of information from previous time steps, enhancing the model's ability to handle varying lengths of dependencies in time series data without the risk of vanishing gradients.

6.3.2.3 Modeling Step

The GRU model was utilized for noise prediction through a two-layer GRU network architecture, comprising two GRU layers and one dense layer. The modeling process is delineated as follows:

1. Input Sequence Encoding:

   Sequential input data is fed into the GRU, which processes one-time step at a time while maintaining a hidden state that captures past information.
2. Gate Mechanisms:

The update and reset gates control how much past information is carried forward and how much new information is incorporated at each time step.

3. Hidden State Update:

   The hidden state is updated based on a combination of the previous hidden state and the current input, regulated by the gates to avoid vanishing gradient problems.

4. Prediction:

   The final hidden state or the sequence of hidden states is used to generate predictions for tasks like sequence classification or time series forecasting.

5. Training:

   The model is trained by minimizing a loss function, typically using backpropagation through time to update the weights.

In the two-layer GRU network employed in this study, the first layer processes historical data and outputs hidden states. These hidden states serve as the input to the second GRU layer, which further processes the information and outputs its hidden states. The second GRU layer is connected to a dense layer, which integrates the information from all neurons in the upper layer to produce the final predicted noise values.

6.3.3 *Bidirectional Long Short-term Memory*

6.3.3.1 Theoretical Basis

Bidirectional Long Short-Term Memory (BiLSTM) networks extend traditional LSTM networks by capturing dependencies in both forward and backward directions within sequence data. LSTMs address the vanishing gradient problem in standard Recurrent Neural Networks (RNNs) using gating mechanisms. BiLSTM further improves this by incorporating contextual information from both past and future states, enabling a more comprehensive understanding of the data. The BiLSTM architecture is depicted in Fig. 6.5.

In this architecture, forward and backward LSTM layers are defined as:

$$h_t^{\text{forward}} = LSTM_{\text{forward}}\left(x_t, h_{t-1}^{\text{forward}}, C_{t-1}^{\text{forward}}\right) \tag{6.6}$$

$$h_t^{\text{backward}} = LSTM_{\text{backward}}\left(x_t, h_{t+1}^{\text{backward}}, C_{t+1}^{\text{backward}}\right) \tag{6.7}$$

BiLSTM combines the forward hidden state h_t^{forward} and the backward hidden state h_t^{backward} to form a more robust representation, allowing the model to utilize both past and future information in making predictions.

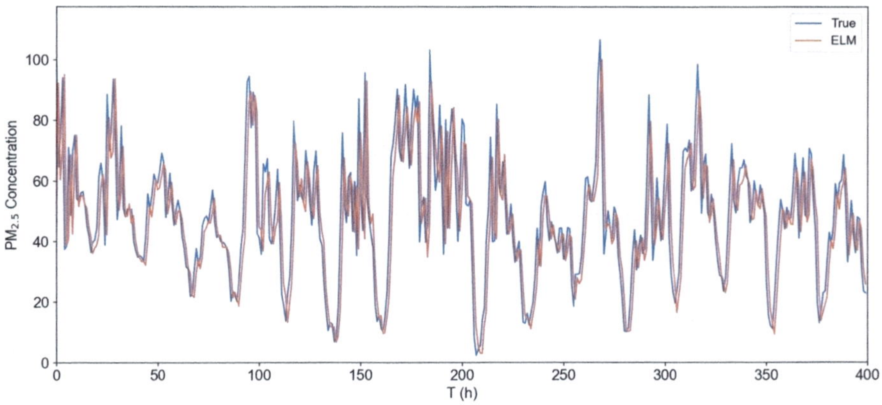

Fig. 6.5 The architecture of the BiLSTM

6.3.3.2 Key Features

Dual-Stream Data Processing: BiLSTM captures bidirectional dependencies that single-directional models like GRU cannot. This bidirectional flow enhances the model's understanding of context from both past and future states, making it ideal for tasks that require comprehensive sequence interpretation.

Improved Memory Retention: With separate forward and backward layers, BiLSTM can retain both short- and long-term information effectively. This ability is particularly useful in tasks requiring the identification of patterns spread across long time series, where conventional LSTM or GRU models may struggle.

6.3.3.3 Modeling Step

1. Input Sequence Processing:
   The input sequence is simultaneously processed in two directions: forward (from past to future) and backward (from future to past), capturing information from both time perspectives.
2. LSTM Cell Operations:
   Each LSTM cell manages an internal memory (cell state) that is updated through forget, input, and output gates, allowing the model to learn long-term dependencies.
3. Bidirectional Hidden States:
   The hidden states of the forward and backward LSTM layers are concatenated, creating a richer representation of the input sequence.
4. Prediction:
   The concatenated hidden states are used to make predictions, leveraging both past and future context information.
5. Training:

The model is trained using backpropagation through time, with weights updated based on the combined forward and backward hidden states.

6.3.4 Deep Extreme Learning Machine

6.3.4.1 Theoretical Basis

The DELM is an extension of the ELM, designed to leverage the hierarchical feature extraction capabilities of deep learning. In contrast to the single-layer architecture of ELM, DELM consists of multiple hidden layers, each of which learns increasingly abstract representations of the input data. The key idea behind DELM is to combine the simplicity and speed of ELM with the powerful feature-learning ability of deep networks. The structure of DELM is shown in Fig. 6.6.

DELM uses stacked ELMs, where each hidden layer's output becomes the input for the next layer. The hidden neurons in each layer are randomly initialized, while the output layer weights are determined through a closed-form solution. This approach provides the benefits of deep learning (hierarchical feature extraction) while maintaining the fast-training process of ELM. The fundamental equation for ELM (previously introduced) applies to each hidden layer in DELM, where the output matrix for each layer becomes the input for the next layer (previously introduced for ELM):

$$T = H\beta \tag{6.8}$$

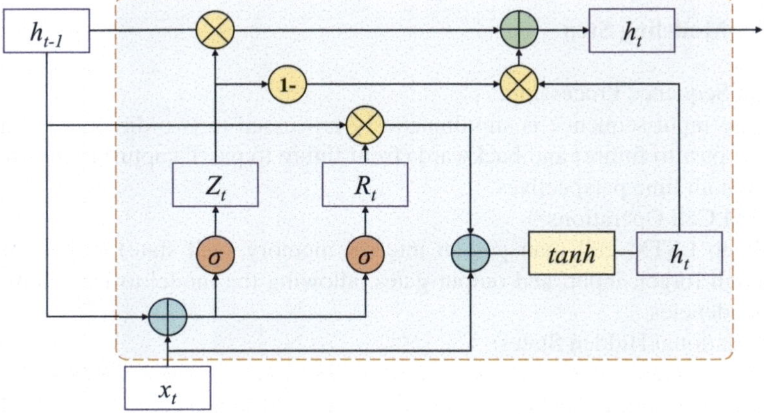

Fig. 6.6 The structure of the DELM

6.3.4.2 Key Features

Adaptive Learning Capabilities: DELM extends the traditional ELM by incorporating a dynamic updating mechanism, allowing the model to adapt to new data over time without retraining from scratch. This adaptability is particularly suited for time series prediction, where evolving patterns need to be captured continuously.

Real-Time Prediction with Online Learning: DELM supports online learning, which enables real-time updating of the model's parameters as new data points become available. This capability makes DELM especially efficient in environments with streaming data or time-dependent inputs.

6.3.4.3 Modeling Step

1. Layer-Wise Initialization:
   Multiple hidden layers are initialized randomly, similar to ELM, but with a deep architecture that allows for more complex feature extraction.
2. Hierarchical Feature Learning:
   Each layer in the deep architecture extracts higher-level features from the input, transforming the data step-by-step through nonlinear activation functions.
3. Output Weight Calculation:
   Similar to standard ELM, the output weights connecting the final hidden layer to the output layer are calculated in one step using a closed-form solution.
4. Model Training:
   The model is trained in a feedforward manner, with the final output weights calculated without backpropagation or iterative tuning of the hidden layers.
5. Prediction:
   For new input, the data is processed through all the hidden layers, and predictions are made based on the learned output weights.

6.3.5 Transformer

6.3.5.1 Theoretical Basis

The Transformer model, introduced by Vaswani et al. in 2017 (Huang et al. 2023; Bentsen et al. 2023), is a deep learning architecture that exclusively uses attention mechanisms to capture dependencies in sequential data. Unlike traditional RNNs or LSTMs, which process sequences step-by-step, the Transformer enables parallel processing by leveraging self-attention to assess relationships across all elements in the input sequence simultaneously. This innovation has significantly impacted fields like natural language processing (NLP) and time-series forecasting. The Transformer architecture is illustrated in Fig. 6.7.

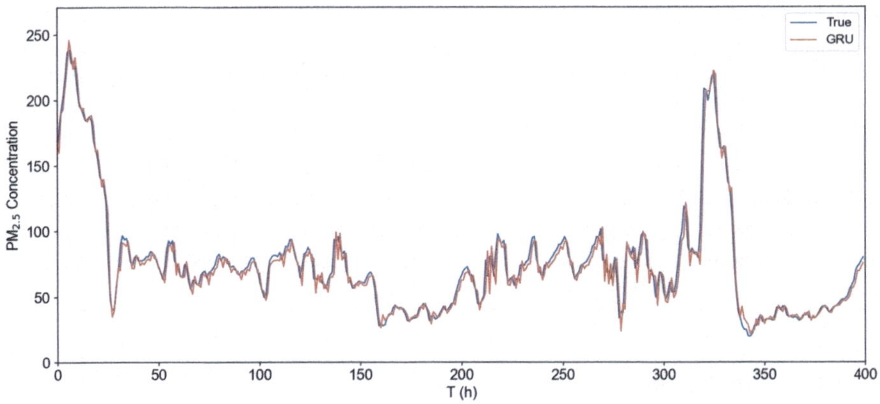

Fig. 6.7 The architecture of the Transformer

The core of the Transformer lies in the self-attention mechanism, which calculates a weighted sum of input elements, enabling the model to focus on the most important parts of the sequence for each position. Multiple attention heads are used across layers to capture diverse relationships within the data. Additionally, position-wise feedforward layers and residual connections are integrated into the architecture to facilitate learning. The self-attention mechanism is defined as:

$$\text{Attention}(Q,K,V) = \text{Softmax}\left(\frac{\bar{Q}K^T}{\sqrt{d}}\right)v \tag{6.9}$$

where, Q, K and V are the query, key, and value matrices, respectively. dk is the dimensionality of the key vectors. The SoftMax function normalizes the attention scores to compute the final weighted sum.

6.3.5.2 Key Features

Self-Attention for Contextual Learning: Transformers rely on a self-attention mechanism that allows the model to focus on relevant parts of the input sequence, regardless of their position. This attention mechanism enables the model to capture long-range dependencies more effectively than traditional RNN-based architectures like GRU or LSTM.

Parallel Processing for Scalability: Unlike sequential models, the Transformer architecture processes all input tokens simultaneously, enabling parallelization. This feature significantly improves the model's scalability, allowing it to handle long sequences with greater computational efficiency, which is advantageous for large-scale time series forecasting.

6.3.5.3 Modeling Step

1. Data Preprocessing:
   Prepare the time series data by normalizing or scaling the input values. Create input-output pairs by forming time windows for sequence prediction, where previous time steps are used to predict future steps.
2. Embedding of Input Sequences:
   Convert the input time series data into embeddings, often by applying positional encodings.
3. Self-Attention Mechanism:
   For each time step, the model computes attention scores that determine how much focus to place on other time steps.
4. Feed-Forward Network:
   After applying self-attention, the resulting values are passed through a feed-forward neural network. This network operates on each position independently, adding non-linearity and improving the model's ability to learn complex patterns in the time series data.
5. Stacking Layers:
   To enhance the model's ability to capture complex dependencies and patterns, multiple layers of self-attention and feed-forward networks are combined. Each layer further refines the learned representations from the previous one.
6. Prediction Layer:
   The final output of the stacked layers is passed through a linear transformation to produce the predicted values for the target time series. This step involves mapping the learned representations back into the original time domain for forecasting future values.

6.4 Probabilistic Forecasting of Air Quality Data

6.4.1 Bayesian Neural Networks

6.4.1.1 Theoretical Basis

The BNN extends traditional neural networks by incorporating probabilistic principles into their architecture, thereby enabling the quantification of uncertainty in predictions. Unlike conventional neural networks that assign fixed values to weights and biases, BNN treats these parameters as probability distributions. This probabilistic approach allows BNN to capture the inherent uncertainty in data and model predictions, which is particularly beneficial in scenarios where understanding the confidence of predictions is crucial.

The foundational concept of BNN is rooted in Bayesian inference, where prior beliefs about the network parameters are updated with observed data to form posterior distributions. By integrating over these distributions during prediction, BNN provides a distribution of possible outcomes rather than single point estimates. This characteristic makes BNN highly effective for probabilistic forecasting, as they provide an understanding of the uncertainty and trustworthiness of the predictions.

A pivotal equation in BNN encapsulates the posterior distribution of the network weights given the data:

$$P(W|D) = \frac{P(D|W)P(W)}{P(D)} \tag{6.10}$$

where, $P(W|D)$ is the posterior distribution of the weights, $P(D|W)$ is the likelihood of the data given the weights, $P(W)$ is the prior distribution of the weights, $P(D)$ is the marginal likelihood of the data.

6.4.1.2 Key Features

Uncertainty Estimation through Bayesian Inference: BNN incorporates Bayesian inference to quantify uncertainty in model predictions. By treating the weights of the network as probability distributions rather than fixed values, BNN provides a robust framework for estimating the uncertainty associated with predictions, which is crucial in applications where risk assessment is vital.

Regularization via Prior Distributions: Incorporating prior distributions in BNN acts as a regularization technique, helping to mitigate overfitting and improve generalization capabilities. This approach allows BNN to leverage prior knowledge about the problem domain, resulting in more reliable predictions in uncertain environments.

6.4.1.3 Modeling Step

1. Model Initialization:
   The neural network architecture is defined, and the weights are initialized as probability distributions rather than fixed values.
2. Forward Propagation with Uncertainty:
   Input data is passed through the network, and weights are sampled from their respective distributions during forward passes.
3. Bayesian Inference:
   Bayes' theorem is applied to update weight distributions as new data becomes available.
4. Posterior Prediction:
   For new inputs, the model outputs a posterior predictive distribution, reflecting the uncertainty in predictions.
5. Training:

Variational inference or Monte Carlo methods are used to approximate the posterior distribution of the weights.

6. Prediction:

   Final predictions are made by averaging multiple forward passes, providing a probabilistic forecast.

6.4.2 *Quantile Recurrent Neural Networks*

6.4.2.1 Theoretical Basis

The QRNN is designed to produce probabilistic forecasts by directly predicting specific quantiles of the target distribution, rather than focusing solely on point estimates. This approach provides a more comprehensive understanding of the potential variability and uncertainty in future observations, which is particularly valuable in time series forecasting where capturing the range of possible outcomes is essential.

QRNN integrates the capabilities of recurrent neural networks (RNN), with quantile regression techniques. By optimizing for quantile loss functions, QRNN learns to estimate the conditional quantiles of the target variable at each time step. This enables the generation of prediction intervals, which can inform decision-making processes by highlighting both the expected value and the associated uncertainty.

A fundamental aspect of QRNN is its ability to handle multiple quantiles simultaneously, allowing for the construction of full predictive distributions. This is achieved by configuring the network to output multiple quantile estimates, each corresponding to a different percentile of the target distribution. As a result, QRNN offers a flexible and robust framework for probabilistic forecasting, accommodating various levels of confidence in the predictions.

$$f_t = \sigma\left(W_f{}^* x_t + b_f\right) \tag{6.11}$$

$$o_t = \sigma\left(W_o{}^* x_t + b_o\right) \tag{6.12}$$

$$c_t = f_t \odot c_{t-1} + \left(1 - f_t\right) \odot W_c{}^* x_t \tag{6.13}$$

$$h_t = o_t \odot c_t \tag{6.14}$$

where, x_t denotes the input at time step t, f_t, o_t, and c_t are the forget gate, output gate, and cell state at time step t, respectively, W_f, W_o, and W_c are the convolutional filters applied to the input x_t to derive the forget gate, output gate, and candidate cell state, b_f, b_o are the bias terms for the forget and output gates, σ denotes the sigmoid activation function, which controls the gating mechanisms, $\odot$ represents element-wise multiplication, h_t is the hidden state at time step t, which is the element-wise product of the output gate o_t and the cell state c_t.

6.4.2.2 Key Features

Quantile Regression for Distributional Predictions: QRNN integrates quantile regression techniques into the recurrent neural network framework, allowing it to predict not only point estimates but also quantiles of the target distribution. This feature enables QRNN to capture the entire distribution of potential outcomes, making it valuable for applications requiring probabilistic forecasts.

Dynamic Temporal Representations: By combining the strengths of both RNN and quantile regression, QRNN can dynamically represent temporal information while maintaining the flexibility to adjust to varying input distributions. This adaptability enhances the model's performance in capturing the nuances of time-dependent data, leading to more accurate probabilistic predictions.

6.4.2.3 Modeling Step

1. Data Normalization:
   The time series data is prepared, and any missing values or anomalies are handled through normalization or imputation.
2. Model Initialization:
   A recurrent neural network architecture, often based on LSTM or GRU, is set up to handle sequential time series data.
3. Quantile Loss Function:
   The network is trained using a quantile loss function that encourages the model to predict specific quantiles of the target distribution.
4. Recurrent Training:
   The model learns temporal dependencies in the data through recurrent layers, allowing it to make accurate predictions over time.
5. Quantile Prediction:
   At each time step, the model predicts multiple quantiles (e.g., lower, median, upper) for the future target distribution.
6. Uncertainty Capture:
   The multiple quantiles provide a range of potential outcomes, giving insight into both the median forecast and extreme scenarios.

6.5 Forecasting Performance Comparison

6.5.1 Evaluation Indicator

6.5.1.1 Deterministic Forecasting

This study employs three widely used statistical error evaluation metrics—mean absolute error (MAE), root mean square error (RMSE), and mean absolute percentage error (MAPE)—to quantitatively assess the model's prediction accuracy. Lower values of these metrics indicate better forecasting performance of the corresponding model. Suppose y_i is the actual value of $PM_{2.5}$ concentration in time i, f_i is the predicted value of the model in time i, and N is the total number of samples in each testing set. The specific formulas for calculating these three-evaluation metrics are provided in Table 6.4.

6.5.1.2 Probabilistic Forecasting

The loss function used for training the neural network combines the Median Absolute Deviation losses for two quantiles. The combination weights for these losses are set to one. The loss function for the quantile regression neural network can be expressed as:

$$\text{loss} = \sum_{\alpha \in [0.5\%, 99.5\%]} \frac{1}{N} \sum_{i=1}^{N} \text{pinball}\left(\alpha, y_i, \hat{y}_{\alpha,i}\right) \tag{6.15}$$

where y_i is the actual observed value for the i-th instance, $\hat{y}_{\alpha,i}$ is the forecasted result for the α-th quantile at instance i, and the pinball loss is defined as follows:

$$\text{pinball}\left(\alpha, y_t, \hat{y}_t(\alpha)\right) = \begin{cases} (1-\alpha)\left(\hat{y}_t(\alpha) - y_t\right), & \text{if } y_t < \hat{y}_t(\alpha) \\ \alpha\left(y_t - \hat{y}_t(\alpha)\right), & \text{if } y_t \geq \hat{y}_t(\alpha) \end{cases} \tag{6.16}$$

where N is the total number of data points, α represents the quantile (e.g., 0.5% or 99.5%), y_i and y_t are the actual observed values for instance iii or time step t, $\hat{y}_{\alpha,i}$ and $\hat{y}_t(\alpha)$ are the forecasted values at quantile α.

PICP measures the proportion of actual values that fall within the predicted intervals (PI). A PICP value closer to the target coverage rate indicates more accurate prediction intervals. This metric helps assess how well the predictive model captures the uncertainty of the data.

MPIW quantifies the average width of the prediction intervals. A smaller MPIW indicates a more precise prediction; however, it should be balanced with the

Table 6.4 The calculation formulas of deterministic forecasting evaluation indicators

Evaluation indicator	Calculation formula
MAE	$\frac{1}{N}\sum_{i=1}^{N}\lvert y_i - f_i\rvert$
RMSE	$\sqrt{\frac{1}{N}\sum_{i=1}^{N}(y_i - f_i)^2}$
MAPE	$\frac{100}{N}\sum_{i=1}^{N}\left\lvert\frac{y_i - f_i}{y_i}\right\rvert$

Table 6.5 The calculation formulas of probabilistic forecasting evaluation indicators

Evaluation indicator	Calculation formula
PICP	$\frac{1}{N}\sum_{i=1}^{N}\mathbb{I}\left(y_i \in [L_i, U_i]\right)$
MPIW	$\frac{1}{N}\sum_{i=1}^{N}(U_i - L_i)$
CWC	(1 − PICP) + MPIW

coverage rate to ensure that the intervals still cover the actual values. Thus, while a narrower MPIW is desirable, it must not come at the expense of PICP.

CWC combines the PICP and MPIW to assess both the accuracy and compactness of the prediction intervals. This metric penalizes the model if the PICP falls below the target coverage rate γ, thus enforcing a trade-off between coverage and interval width. The formulas used to compute these three metrics are presented in Table 6.5.

These metrics are widely used to evaluate the quality of interval forecasts in probabilistic prediction. The pinball loss is a standard loss function for quantile regression, making it suitable for handling predictions across different quantiles. PICP and MPIW provide insights into the coverage and precision of the prediction intervals, while CWC offers a comprehensive evaluation by balancing both accuracy and interval width. In practical applications, the objective is to achieve both high PICP and low MPIW, ensuring that the intervals are accurate and concise.

6.5.2 Deterministic Forecasting Performance

The deterministic forecasting results of each model on the two datasets are shown in Figs. 6.8, 6.9, 6.10, 6.11, 6.12, 6.13, 6.14, 6.15, 6.16 and 6.17.

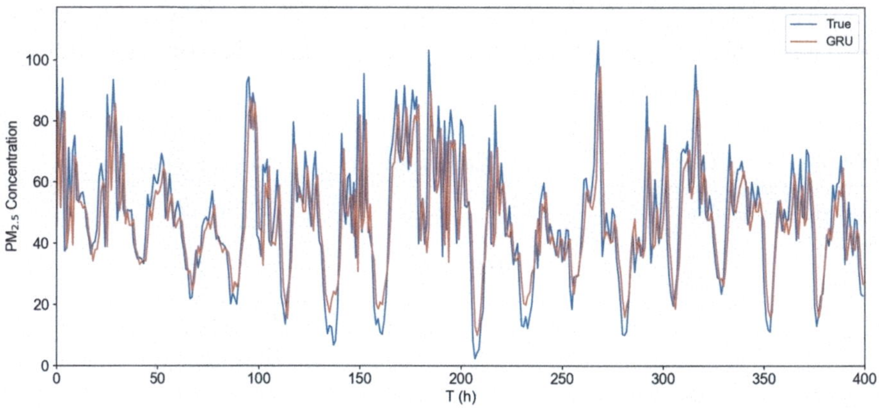

Fig. 6.8 ELM forecasting results of the data #1

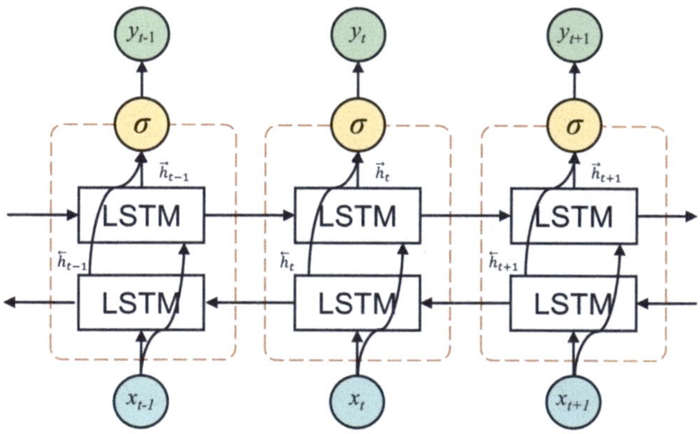

Fig. 6.9 ELM forecasting results of the data #2

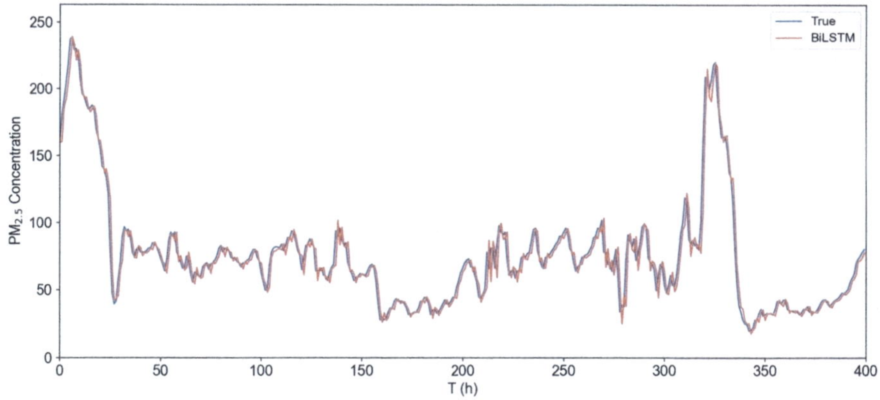

Fig. 6.10 GRU forecasting results of the data #1

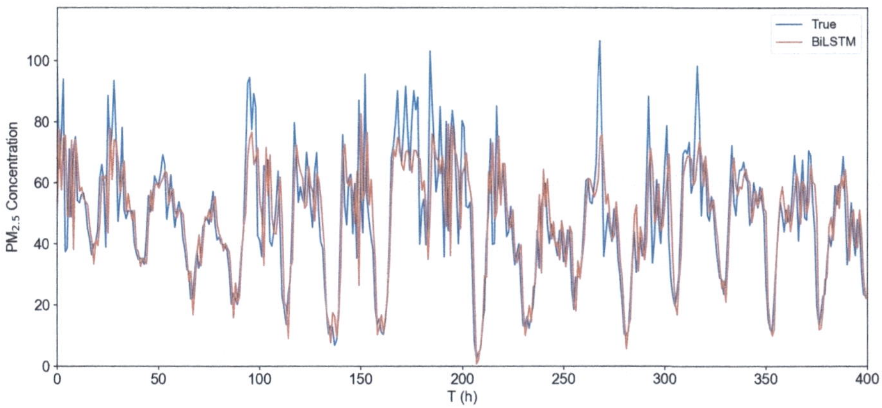

Fig. 6.11 GRU forecasting results of the data #2

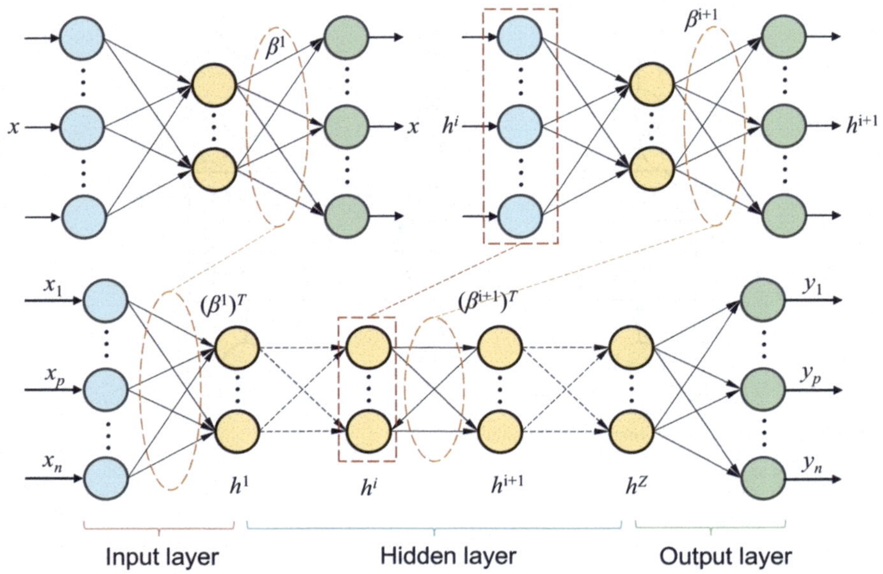

Fig. 6.12 BiLSTM forecasting results of the data #1

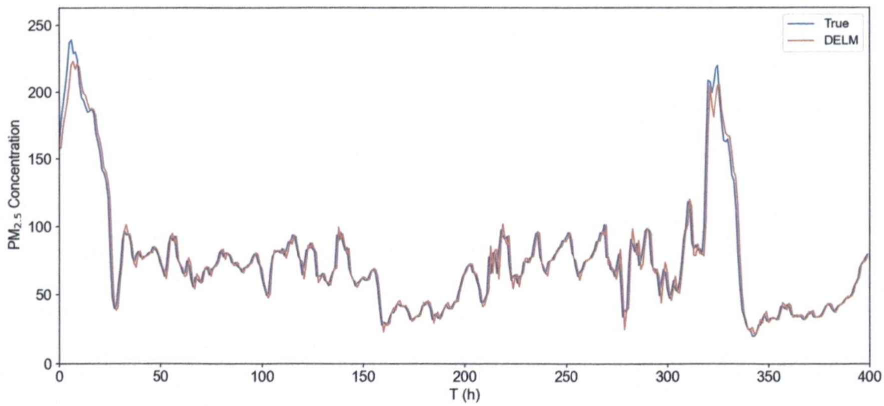

Fig. 6.13 BiLSTM forecasting results of the data #2

In data #1, the performance of the five forecasting models—ELM, GRU, BiLSTM, DELM, and Transformer—was evaluated using three key error metrics: MAE, RMSE, and MAPE. Among these models, BiLSTM demonstrated the best overall accuracy, achieving the lowest values for both MAE (5.47 μg/m^3) and MAPE (7.80%), indicating that it consistently provides reliable predictions of $PM_{2.5}$ levels.

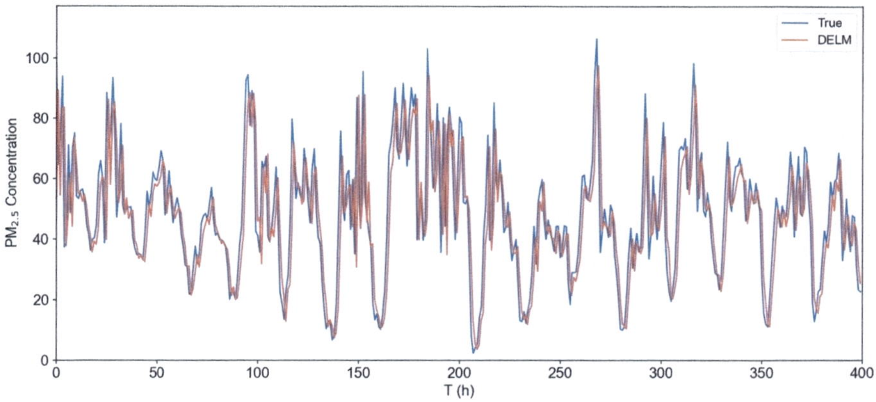

Fig. 6.14 DELM forecasting results of the data #1

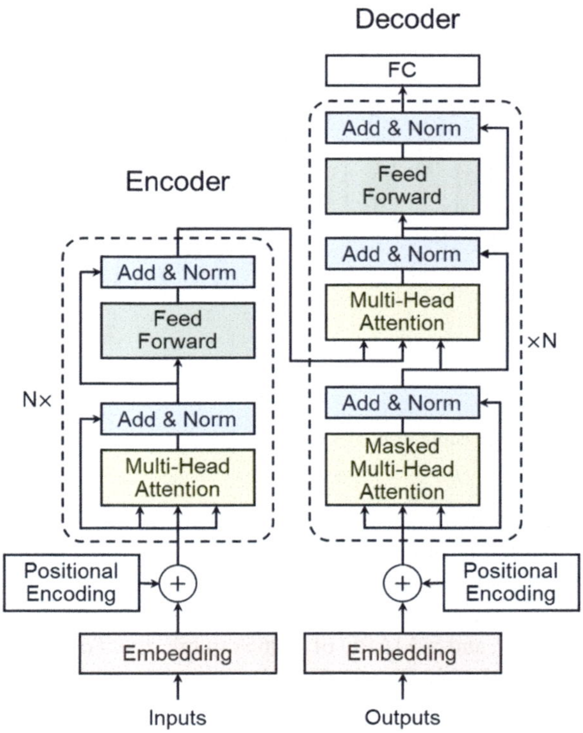

Fig. 6.15 DELM forecasting results of the data #2

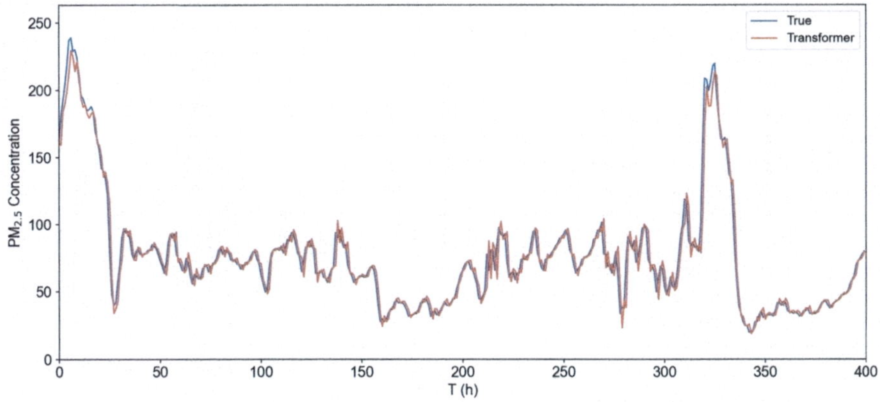

Fig. 6.16 Transformer forecasting results of the data #1

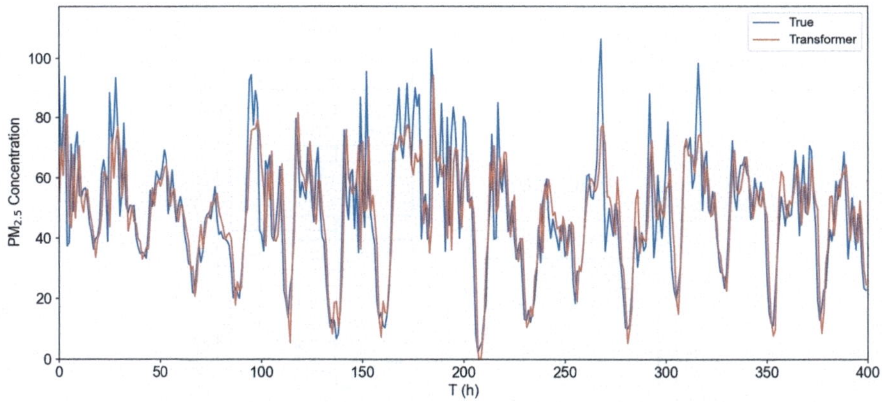

Fig. 6.17 Transformer forecasting results of the data #2

In contrast, ELM exhibited the highest MAE (6.59 μg/m^3) and RMSE (10.29 μg/m^3), suggesting that it tends to produce larger errors compared to the other models. This is particularly evident in its MAPE, which highlights its less reliable performance. GRU and the Transformer models also performed well, with GRU yielding a MAE of 5.61 μg/m^3 and a MAPE of 8.06%, while the Transformer achieved a MAE of 5.39 μg/m^3 and a MAPE of 7.58%. Both models indicate a strong ability to capture the underlying data patterns, contributing to their performance.

DELM, while slightly less effective than BiLSTM and GRU, still provided reasonable predictions with a MAE of 5.89 μg/m^3 and a MAPE of 7.98%. The variability in model performance underscores the importance of selecting appropriate forecasting techniques based on specific dataset characteristics. Overall, the results suggest that BiLSTM, followed by the Transformer and GRU models, are the most suitable choices for accurate $PM_{2.5}$ predictions, particularly in contexts where both absolute and percentage errors are critical for decision-making.

In analyzing the forecasting performance of the models on dataset 2, we observe a general trend of higher error metrics compared to dataset 1, indicating challenges in accurately predicting $PM_{2.5}$ levels. BiLSTM emerges as the most effective model, achieving the lowest Mean Absolute Error (MAE) of 9.43 μg/m^3 and a Mean Absolute Percentage Error (MAPE) of 21.92%. This suggests that BiLSTM is capable of capturing the underlying patterns in this dataset, evidenced by its favorable Root Mean Square Error (RMSE) of 13.14 μg/m^3, reflecting a lower degree of prediction variance.

In contrast, ELM demonstrates the highest error metrics, with an MAE of 10.88 μg/m^3 and an RMSE of 14.92 μg/m^3, indicating a tendency to produce larger forecasting errors. Its MAPE of 26.05% further highlights its inadequacy in accurately modeling the data. GRU and DELM also present reasonable performances, with GRU recording an MAE of 9.91 μg/m^3 and a MAPE of 25.92%, while DELM shows a MAE of 10.35 μg/m^3 and a MAPE of 25.01%. Although these models perform adequately, they do not achieve the accuracy levels of BiLSTM or the Transformer.

The Transformer model also exhibits competitive results, with an MAE of 9.2929 μg/m^3 and the lowest RMSE of 13.02 μg/m^3, paired with a MAPE of 22.62%. This performance indicates its reliability relative to other models in this dataset. Overall, while all models reflect increased error rates in dataset 2, BiLSTM and the Transformer stand out as the most promising options for accurate $PM_{2.5}$ predictions, emphasizing the need for further refinement and potential adjustments to enhance forecasting accuracy given the dataset's complexity.

The error evaluations presented in Tables 6.6 and 6.7 provide a comprehensive comparison of the forecasting results for $PM_{2.5}$ levels in Changsha and Seoul. Notably, the performance metrics—MAE, RMSE, and MAPE—reveal significant differences in model effectiveness across the two datasets. Error Analysis Plot of the data #1 on both datasets is shown in Figs. 6.18 and 6.19.

In Changsha, the BiLSTM model outperforms the other models, achieving the lowest MAE of 5.47 μg/m^3, RMSE of 8.33 μg/m^3, and MAPE of 7.79%. This indicates that BiLSTM is particularly adept at accurately predicting $PM_{2.5}$ levels in this dataset. The Transformer model closely follows, with an MAE of 5.39 μg/m^3 and a MAPE of 7.58%, suggesting its strong reliability as well.

Table 6.6 The error evaluation of the forecasting results in Changsha

Model	MAE (μg/m^3)	RMSE (μg/m^3)	MAPE (%)
ELM	6.59	10.29	9.37
GRU	5.61	8.34	8.06
BiLSTM	5.47	8.33	7.80
DELM	5.89	9.35	7.99
Transformer	5.39	8.36	7.58

Table 6.7 The error evaluation of the forecasting results in Seoul

Model	MAE (μg/m³)	RMSE (μg/m³)	MAPE (%)
ELM	10.88	14.92	26.05
GRU	9.91	13.95	25.92
BiLSTM	9.43	13.14	21.92
DELM	10.35	14.37	25.01
Transformer	9.29	13.02	22.62

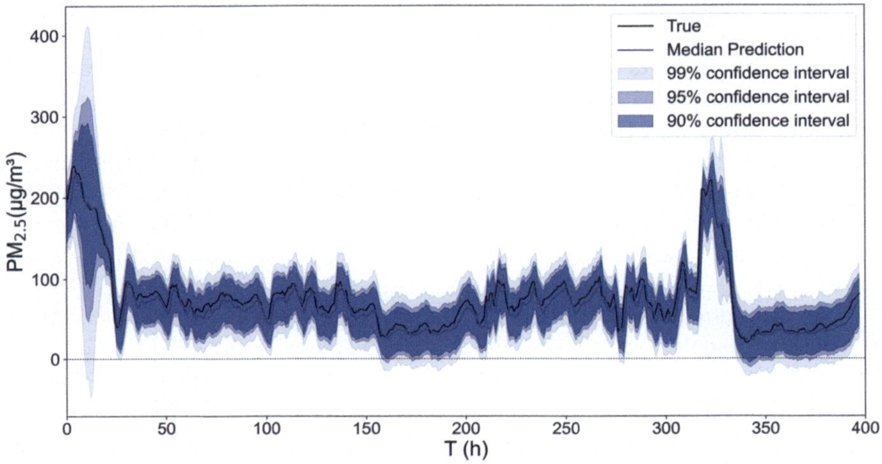

Fig. 6.18 Error Analysis Plot of the data #1

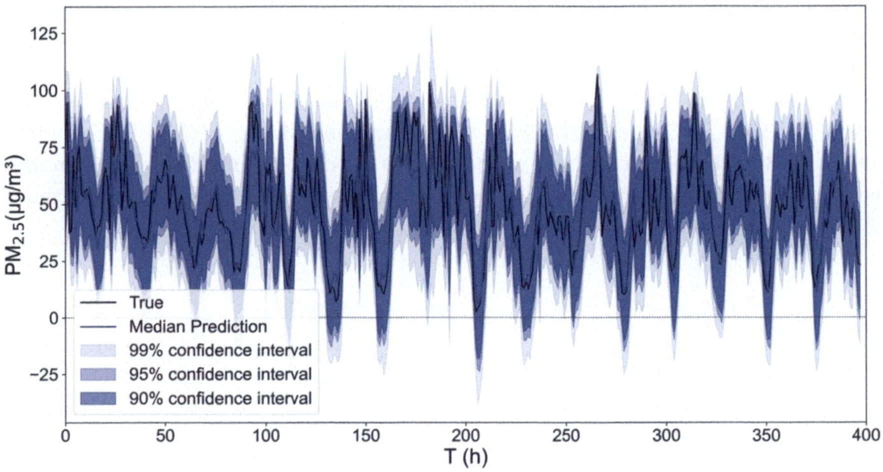

Fig. 6.19 Error Analysis Plot of the data #2

Conversely, in Seoul, all models exhibit higher error metrics compared to their performance in Changsha, indicating a more challenging forecasting environment. ELM shows the highest error rates across all metrics, with an MAE of 10.8 μg/m^3 and a MAPE of 26.1%. This stark increase in errors highlights the model's inadequacy in capturing the complexities of the Seoul dataset. GRU also presents significant error values, with an MAE of 9.91 μg/m^3 and a MAPE of 25.9%.

BiLSTM remains the best-performing model in Seoul as well, but its MAE of 9.42 μg/m^3 and MAPE of 21.9% are notably higher than in Changsha, indicating that while it retains its effectiveness, the increase in error metrics reflects the dataset's greater variability or complexity. The Transformer model again demonstrates competitive performance with an MAE of 9.30 μg/m^3 and a MAPE of 22.6%, suggesting it is a reliable option even in differing conditions.

Overall, the comparison highlights that while BiLSTM consistently performs well across both datasets, the forecasting accuracy diminishes in Seoul. The increased error metrics in Seoul compared to Changsha emphasize the need for tailored modeling approaches that account for the unique characteristics of each dataset. This analysis underscores the importance of model selection and adaptation in achieving reliable air quality forecasts across different geographical contexts.

6.5.3 Probabilistic Forecasting Performance

The probabilistic forecasting results of each model on the two datasets are shown in Figs. 6.20, 6.21, 6.22, and 6.23. Comparison of scatter plots of various deterministic and probabilistic prediction models on both the datasets are shown in Figs. 6.24 and 6.25.

Table 6.8 presents the error evaluation metrics for forecasting $PM_{2.5}$ concentrations in Changsha using two different models: BNN and QRNN. The metrics considered are MAE, RMSE, and MAPE. The median prediction errors for the two datasets for a given time interval are shown in Figs. 6.26 and 6.27. Comparison of the evaluation metrics for the various models on the two datasets is shown in Figs. 6.28 and 6.29.

The QRNN model demonstrates superior performance compared to the BNN model across all three metrics. Specifically, QRNN achieves a 28.6% reduction in MAE, a 12.3% decrease in RMSE, and a 31.6% decline in MAPE relative to BNN. These improvements indicate that QRNN provides more accurate and reliable predictions of $PM_{2.5}$ levels in Changsha, likely due to its enhanced ability to capture temporal dependencies and complex patterns within the data.

Table 6.9 outlines the error evaluation metrics for forecasting $PM_{2.5}$ concentrations in Seoul using the same models. In contrast to the results observed in Changsha, the performance improvement of QRNN over BNN in Seoul is marginal. QRNN exhibits a slight reduction in MAE (approximately 1.4%) and MAPE (around 8.3%) compared to BNN. However, QRNN shows a minor increase in RMSE (about 1.1%) relative to BNN. This negligible difference suggests that while QRNN maintains its

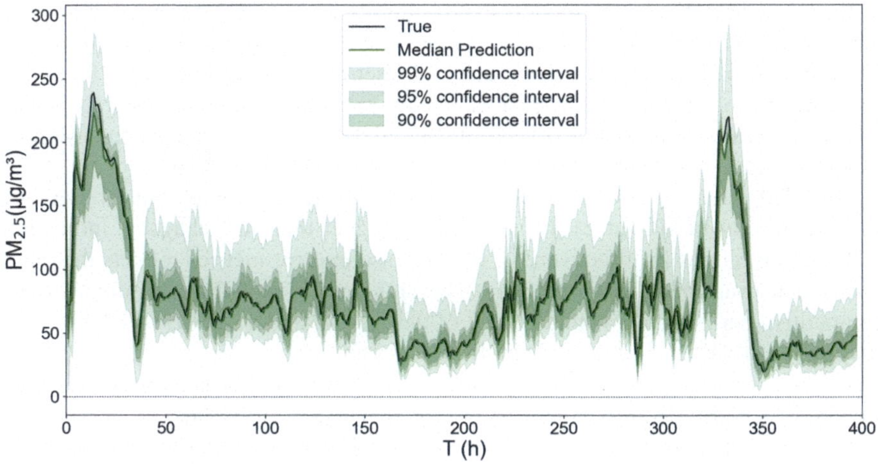

Fig. 6.20 BNN forecasting results of the data #1

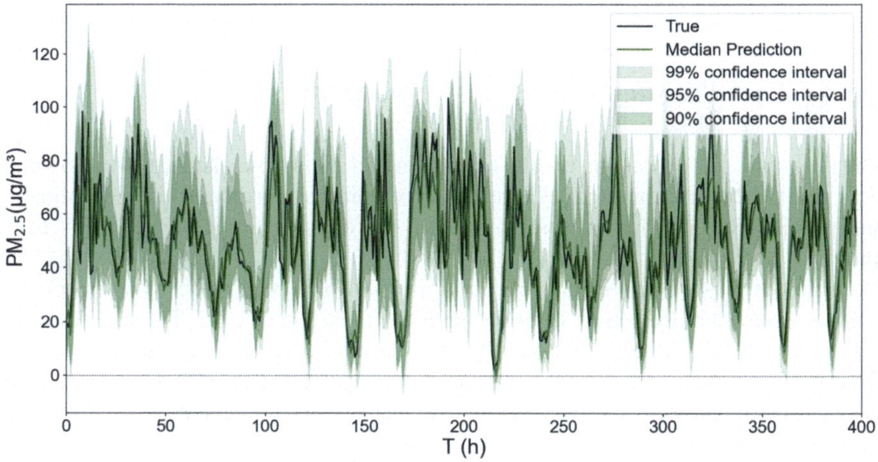

Fig. 6.21 BNN forecasting results of the data #2

predictive capability, the advantage over BNN is less pronounced in the Seoul dataset. The visualization of BNN and QRNN's predictions within 90% confidence intervals for the two sets of data in a given time interval is shown in Figs. 6.30 and 6.31.

The evaluation of prediction interval metrics between Changsha and Seoul demonstrates significant differences in model performance across regions. Tables 6.10 and 6.11 present the Prediction Interval Coverage and Width Metrics for Changsha and Seoul. In Changsha, QRNN consistently outperforms BNN in terms of

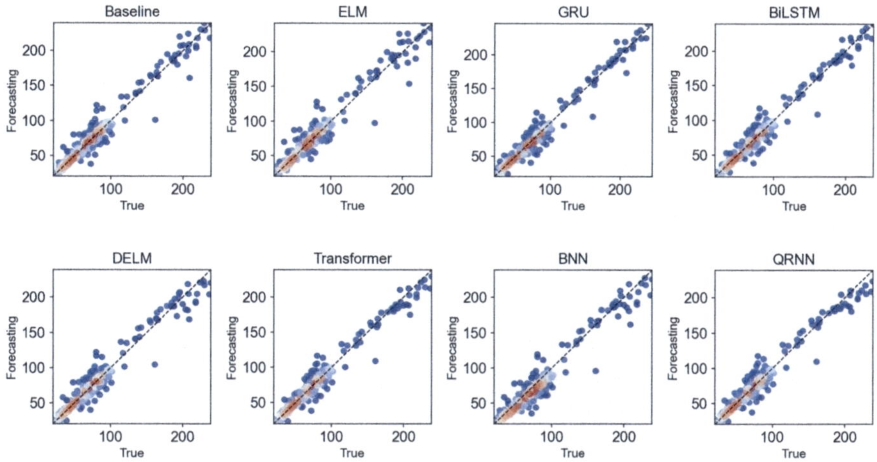

Fig. 6.22 QRNN forecasting results of the data #1

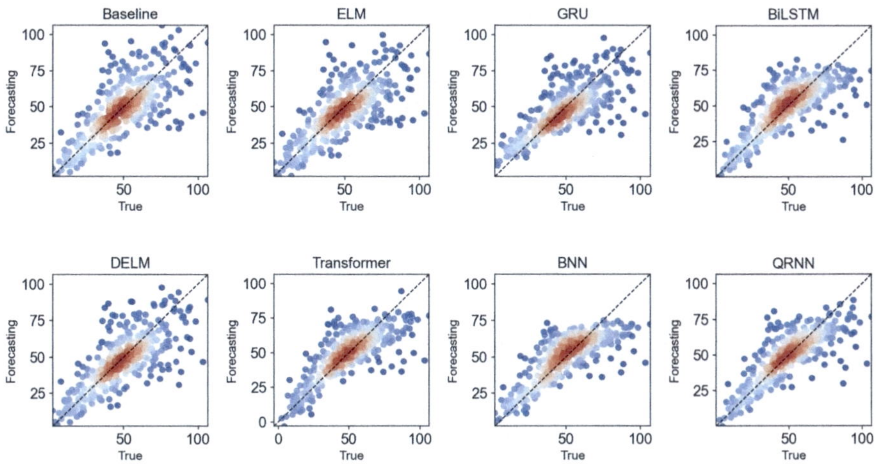

Fig. 6.23 QRNN forecasting results of the data #2

prediction interval efficiency. QRNN produces narrower prediction intervals (lower MPIW) while maintaining reasonable coverage probabilities (PICP). For example, at the 90% interval, QRNN achieves a significantly lower MPIW of 26.22 compared to BNN's 58.10, indicating more compact and precise predictions. Additionally, the lower Coverage Width-based Criterion (CWC) values of QRNN further confirm its ability to balance interval width and coverage in Changsha, making it a more

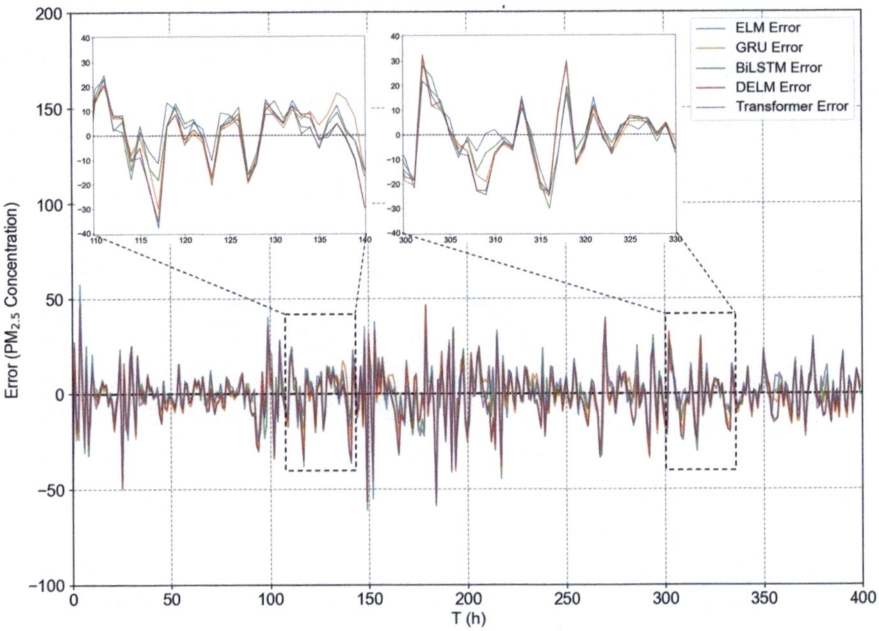

Fig. 6.24 Scatter Plot Comparison of Various Models for the data #1

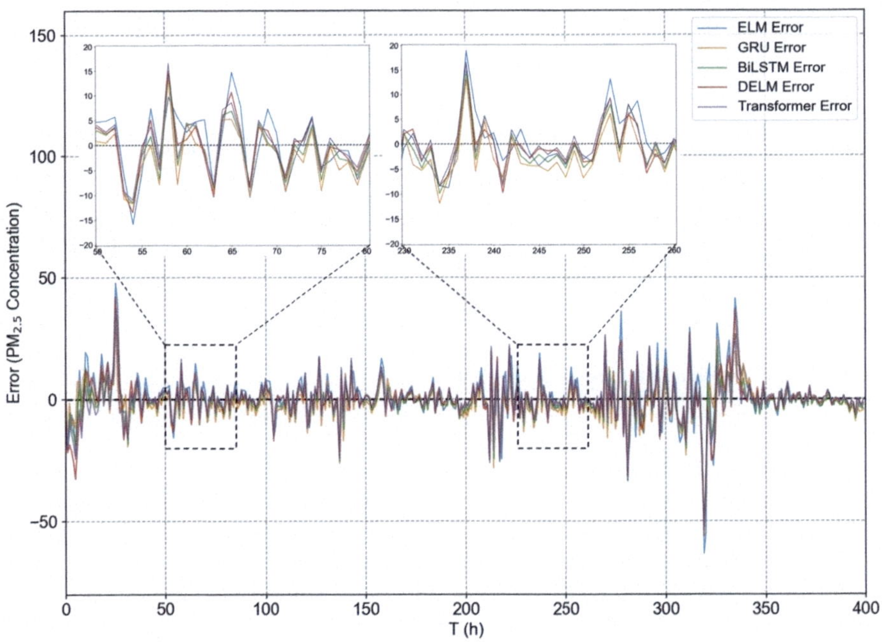

Fig. 6.25 Scatter Plot Comparison of Various Models for the data #2

Table 6.8 Error evaluation of forecasting results in Changsha

Model	MAE (μg/m³)	RMSE (μg/m³)	MAPE (%)
BNN	8.35	11.26	11.85
QRNN	5.99	9.91	8.10

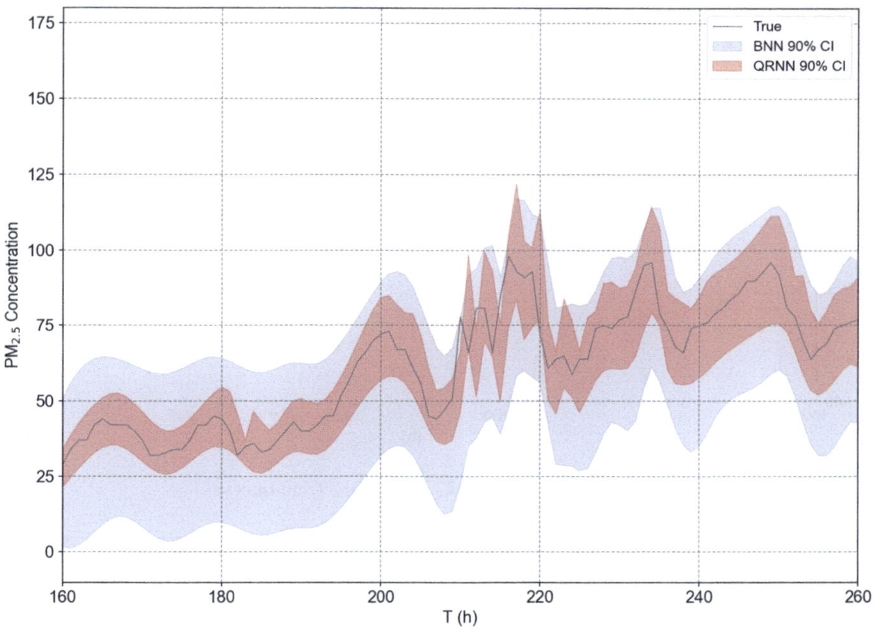

Fig. 6.26 Median Prediction Error of the data #1 in Specific Time Interval

efficient model in this region. Comparison of CWC values for different models in different regions is shown in Fig. 6.32.

In contrast, the performance gap between QRNN and BNN diminishes in Seoul, where environmental factors may contribute to more volatile $PM_{2.5}$ data. In Seoul, BNN exhibits higher coverage probabilities at all confidence intervals, with a PICP of 0.920 for the 90% interval compared to QRNN's 0.857. While QRNN continues to produce narrower intervals in Seoul, the reduction in coverage probability suggests that QRNN's intervals may be too narrow to capture sufficient uncertainty in the data. This is particularly evident at lower confidence levels, where the trade-off between coverage and interval width becomes more pronounced.

The geographical differences between Changsha and Seoul underscore the importance of context-specific model evaluation. In Changsha, QRNN's ability to deliver tighter and more precise intervals without sacrificing too much coverage makes it a preferable choice. However, in Seoul's noisier environment, BNN offers

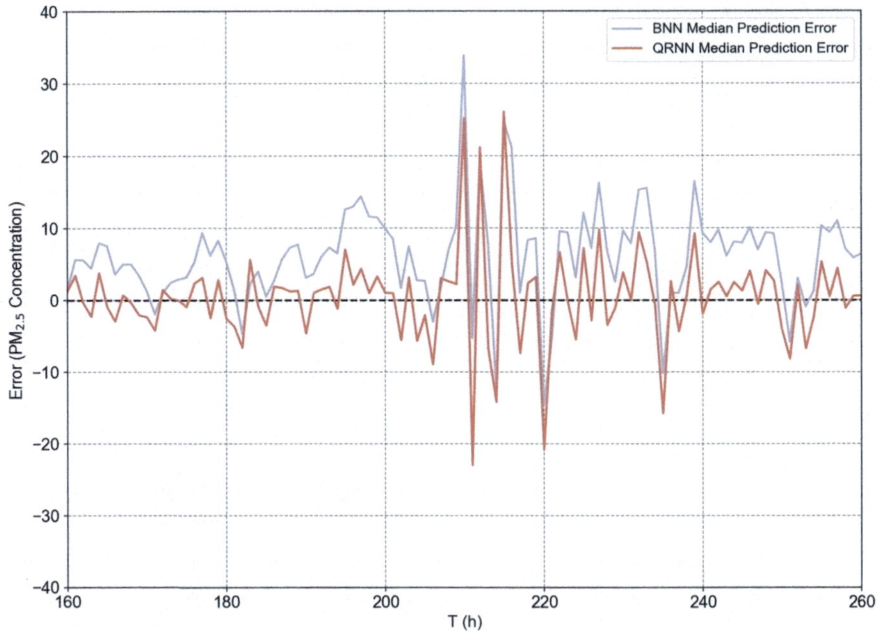

Fig. 6.27 Median Prediction Error of the data #2 in Specific Time Interval

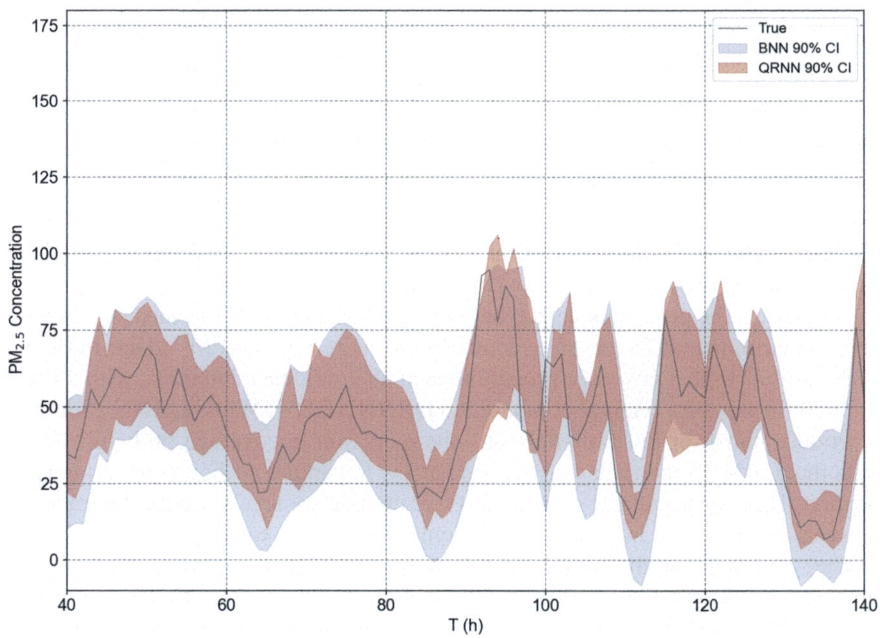

Fig. 6.28 Comparison of Evaluation Metrics for Various Models in data #1

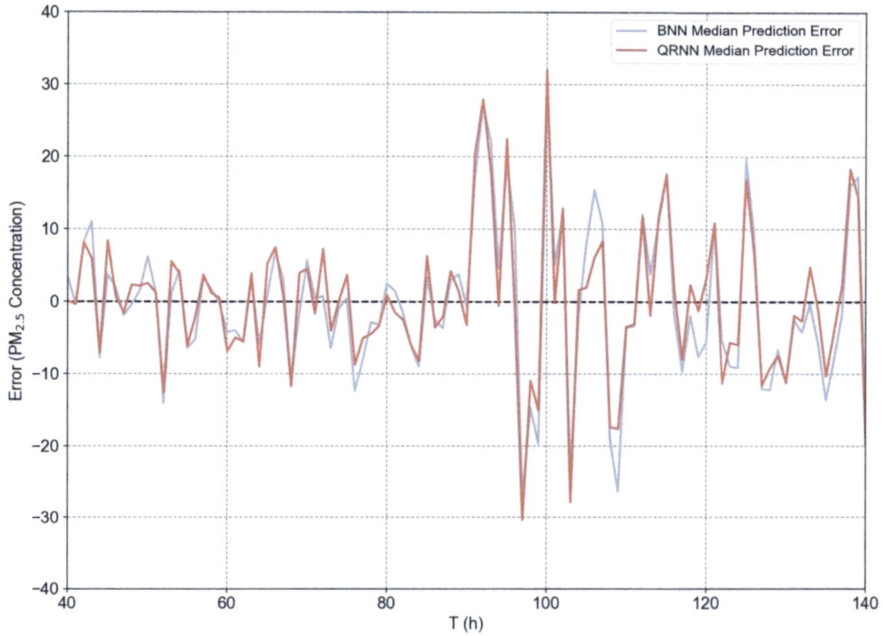

Fig. 6.29 Comparison of Evaluation Metrics for Various Models in data #2

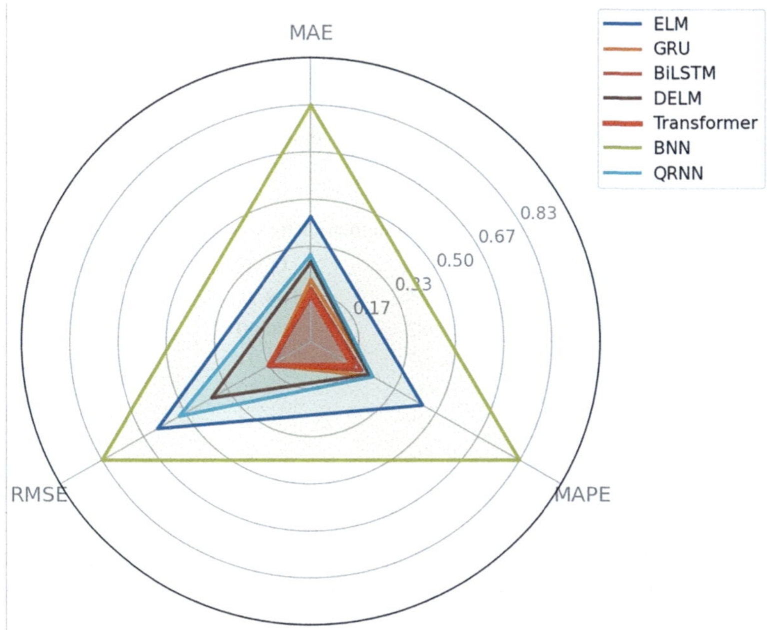

Fig. 6.30 BNN vs. QRNN Predictions of the data #1 within 90% Confidence Interval in Specific Time Interval

Table 6.9 Error evaluation of forecasting results in Seoul

Model	MAE (μg/m³)	RMSE (μg/m³)	MAPE (%)
BNN	9.11	12.49	22.99
QRNN	8.94	12.64	21.17

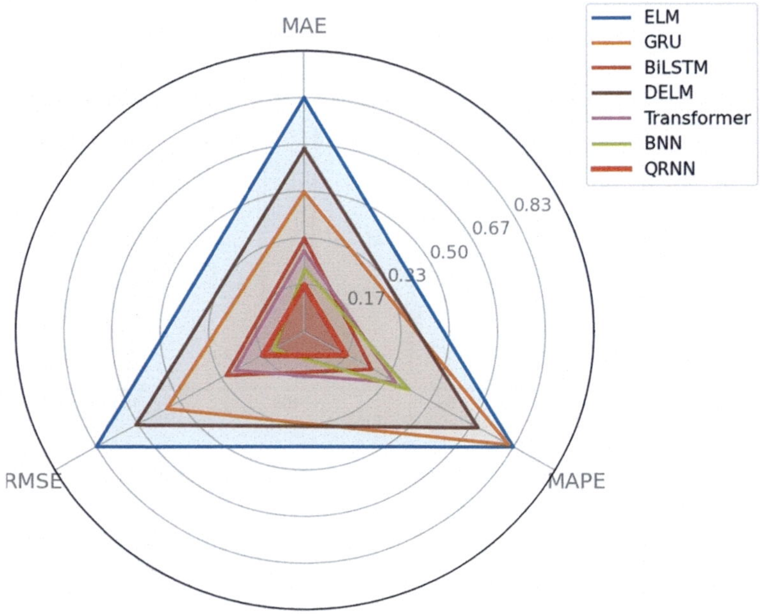

Fig. 6.31 BNN vs. QRNN Predictions of the data #2 within 90% Confidence Interval in Specific Time Interval

Table 6.10 Prediction interval coverage and width metrics for Changsha

Model	Interval	PICP	CWC	MPIW
BNN	90%	0.975	58.71	58.10
	95%	0.982	70.60	70.25
	99%	0.997	95.94	95.81
QRNN	90%	0.874	28.59	26.23
	95%	0.925	37.34	36.03
	99%	0.997	78.88	78.87

a more reliable performance in terms of coverage, albeit with wider prediction intervals. These findings indicate that model selection should be adaptive, based on the data characteristics and volatility of the target region.

Table 6.11 Prediction interval coverage and width metrics for Seoul

Model	Interval	PICP	CWC	MPIW
BNN	90%	0.920	45.10	43.70
	95%	0.955	52.97	52.23
	99%	0.987	69.91	69.67
QRNN	90%	0.857	35.43	32.79
	95%	0.902	41.74	40.18
	99%	0.975	59.32	59.01

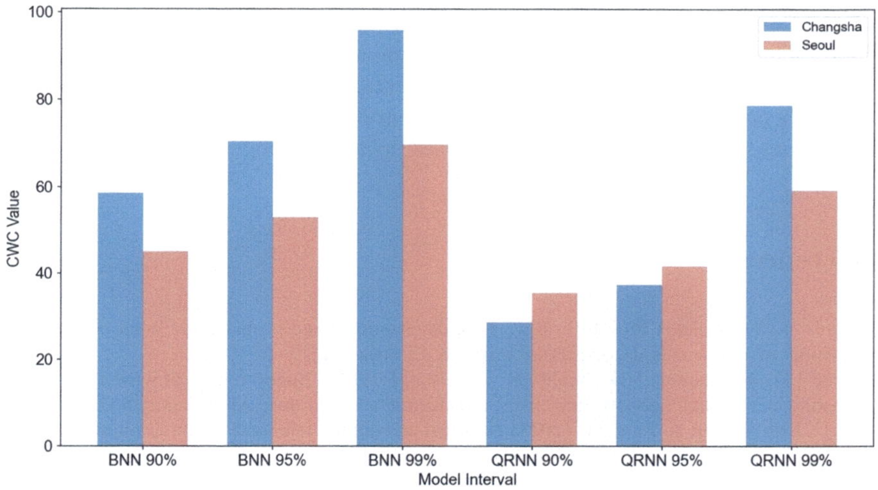

Fig. 6.32 Comparison of CWC Values for Different Models

Given the results, QRNN is well-suited for environments with more structured or stable data patterns, like Changsha. It provides precise and compact prediction intervals, which are crucial in applications where narrow bounds are necessary for accurate air quality control. On the other hand, in regions with more erratic data patterns such as Seoul, BNN's higher coverage reliability makes it a safer option, especially when the goal is to ensure that actual $PM_{2.5}$ concentrations fall within the predicted intervals.

6.6 Conclusions

The comprehensive evaluation of deterministic and probabilistic forecasting models for PM2.5 concentrations across datasets from Changsha and Seoul provides valuable insights into their performance. Among the deterministic models, the BiLSTM and Transformer consistently outperformed others, achieving the lowest MAE, RMSE, and MAPE metrics across both datasets. In contrast, the ELM model exhibited the highest error rates, highlighting its limitations in capturing the data's complexities. In the probabilistic forecasting domain, the QRNN demonstrated superior performance over the BNN in Changsha, offering more accurate prediction intervals with reasonable coverage probabilities. However, in Seoul, the performance gap between the two models narrowed, with the BNN displaying greater reliability in coverage, despite generating wider intervals.

These findings emphasize the importance of adapting forecasting models to the unique characteristics of each dataset and its environmental context. As such, model selection should be context-sensitive, as demonstrated by the differing requirements of the Changsha and Seoul datasets. The continuous refinement of forecasting models and adaptive strategies will be crucial for enhancing the accuracy and reliability of PM2.5 predictions. This, in turn, will contribute to more effective air quality management and decision-making processes in various geographical settings.

References

Bazionis IK, Georgilakis PS (2021) Review of deterministic and probabilistic wind power forecasting: models, methods, and future research. Electricity 2:13–47

Bentsen LØ, Warakagoda ND, Stenbro R, Engelstad P (2023) Spatio-temporal wind speed forecasting using graph networks and novel transformer architectures. Appl Energy 333:120565

Chen C, Liu Y, Chen L, Zhang C (2023) Bidirectional spatial-temporal adaptive transformer for urban traffic flow forecasting. IEEE Trans Neural Netw Learn Syst 34(10):6913–6925

Darekar A, Reddy AA (2017) Predicting market price of soybean in Major India studies through ARIMA model

Duan Z, Liu H, Han F, Li Y (2018) Big multi-step wind speed forecasting model based on secondary decomposition, ensemble method and error correction algorithm. Energy Convers Manag 156:525–541

Elbaz K, Hoteit I, Shaban WM, Shen S-L (2023) Spatiotemporal air quality forecasting and health risk assessment over smart city of NEOM. Chemosphere 313:137636

Fang Y, Liu H (2023) Probabilistic concentration prediction of PM2.5 in subway stations based on multi-resolution elastic-gated attention mechanism and Gaussian mixture model. J Cent South Univ 30:2818–2832

Fang W, Zhu R, Lin JC-W (2023) An air quality prediction model based on improved vanilla LSTM with multichannel input and multiroute output. Expert Syst Appl 211:118422

Guo Z, Yang C, Wang D, Liu H (2023) A novel deep learning model integrating CNN and GRU to predict particulate matter concentrations. Process Saf Environ Prot 173:604–613

Huang B, Dou H, Luo Y, Li J, Wang J, Zhou T (2023) Adaptive spatiotemporal transformer graph network for traffic flow forecasting by IoT loop detectors. IEEE Internet Things J 10:1642–1653

Jia J, Yuan S, Shi Y, Wen J, Pang X, Zeng J (2022) Improved sparrow search algorithm optimization deep extreme learning machine for lithium-ion battery state-of-health prediction. iScience 25:103988

Kim M, SankaraRao B, Kang O, Kim J, Yoo C (2012) Monitoring and prediction of indoor air quality (IAQ) in subway or metro systems using season dependent models. Energ Buildings 46:48–55

Li W, Fu H, Han Z, Zhang X, Jin H (2022) Intelligent tool wear prediction based on informer encoder and stacked bidirectional gated recurrent unit. Robot Comput Integr Manuf 77:102368

Li D, Tan Y, Zhang Y, Miao S, He S (2023) Probabilistic forecasting method for mid-term hourly load time series based on an improved temporal fusion transformer model. Int J Electr Power Energy Syst 146:108743

Liu H, Yang C, Huang M, Wang D, Yoo C (2018) Modeling of subway indoor air quality using Gaussian process regression. J Hazard Mater 359:266–273

Liu H, Yin S, Chen C, Duan Z (2020) Data multi-scale decomposition strategies for air pollution forecasting: a comprehensive review. J Clean Prod 277:124023

Luo L, Dong J, Kong W, Lu Y, Zhang Q (2024) Short-term probabilistic load forecasting using quantile regression neural network with accumulated hidden layer connection structure. IEEE Trans Industr Inform 20:5818–5828

Ma D, Guo Y, Ma S (2021) Short-term Subway passenger flow prediction based on GCN-BiLSTM. IOP Conf Ser Earth Environ Sci 693:012005

Mi X, Liu H, Li Y (2017) Wind speed forecasting method using wavelet, extreme learning machine and outlier correction algorithm. Energy Convers Manag 151:709–722

Nowotarski J, Weron R (2018) Recent advances in electricity price forecasting: a review of probabilistic forecasting. Renew Sust Energ Rev 81:1548–1568

Ren S, Wang X, Zhou X, Zhou Y (2023) A novel hybrid model for stock price forecasting integrating encoder Forest and informer. Expert Syst Appl 234:121080

Son Y-S, Jeon J-S, Lee HJ, Ryu I-C, Kim J-C (2014) Installation of platform screen doors and their impact on indoor air quality: Seoul subway trains. J Air Waste Manage Assoc 64:1054–1061

Su M, Liu H, Yu C, Duan Z (2023) A novel AQI forecasting method based on fusing temporal correlation forecasting with spatial correlation forecasting. Atmos Pollut Res 14:101717

Tan J, Liu H, Li Y, Yin S, Yu C (2022) A new ensemble spatio-temporal PM2.5 prediction method based on graph attention recursive networks and reinforcement learning. Chaos, Solitons Fractals 162:112405

Tissera MD, McDonnell MD (2016) Deep extreme learning machines: supervised autoencoding architecture for classification. Neurocomputing 174:42–49

Wang B, Liu X, Chi M, Li Y (2024) Bayesian network based probabilistic weighted high-order fuzzy time series forecasting. Expert Syst Appl 237:121430

Wu C, He H, Song R, Zhu X, Peng Z, Fu Q, Pan J (2023) A hybrid deep learning model for regional O3 and NO2 concentrations prediction based on spatiotemporal dependencies in air quality monitoring network. Environ Pollut 320:121075

Yin S, Liu H, Duan Z (2021) Hourly PM2.5 concentration multi-step forecasting method based on extreme learning machine, boosting algorithm and error correction model. Digit Signal Process 118:103221

Zhang Y, Wang J, Wang X (2014) Review on probabilistic forecasting of wind power generation. Renew Sust Energ Rev 32:255–270

Chapter 7
Data Interpolation in Air Quality Monitoring

Abstract This chapter provides a comprehensive overview of data interpolation techniques in air quality monitoring, focusing on two key dimensions: temporal and spatial interpolation. In the domain of temporal interpolation, it details the principles and model construction for four widely used methods: linear interpolation, polynomial interpolation, spline interpolation, and interpolation based on statistical models. For spatial interpolation, it explores the theoretical foundations and model design of four classic approaches: nearest neighbor interpolation, inverse distance weighting interpolation, Kriging interpolation, and radial basis function interpolation. To conclude, the chapter presents experimental comparisons of the four temporal and four spatial interpolation methods, comparing the characteristics, performance, and application scenarios of different methods. Additionally, the distinct characteristics and appropriate application scenarios between temporal interpolation and spatial interpolation are highlighted in the chapter.

7.1 Introduction

Air quality is crucial to people's health and quality of life. To improve air quality, continuously or periodically measuring pollutants in the atmosphere through scientific is significant, which is air quality monitoring. In air quality monitoring, data interpolation is an important technical means used to estimate and predict air quality conditions in unknown areas.

In the actual environment, Air quality monitoring stations are often unevenly distributed and limited, unable to cover all regions. Through data interpolation techniques, we can utilize known monitoring data from nearby stations to estimate and predict the air quality conditions in areas without monitoring stations, thereby filling in the monitoring gaps and making the air quality data more spatially continuous and complete (Liu et al. 2018). In addition, Data interpolation methods not only consider the distance relationship between known monitoring points and interpolation points, but also take into account the spatial correlation between known

H. Liu et al., *Data Science in Air Quality Monitoring*, Engineering Applications of Computational Methods 23, https://doi.org/10.1007/978-981-96-5777-3_7

monitoring points (Wang et al. 2023). This comprehensive consideration makes the interpolation results more accurate and reliable and can more truly reflect the regional air quality conditions.

Data interpolation can be categorized based on the characteristics of the interpolation methods themselves and their application scenarios, such as linear interpolation, polynomial interpolation, Lagrange interpolation, and piecewise interpolation. However, the most classical categorization is according to time and space (Roszkowiak et al. 2017). Data interpolation can be mainly divided into two major categories: temporal interpolation and spatial interpolation.

- Temporal interpolation is a technique used to fill in missing time points or resample time series data. Its primary objective is to ensure that the time series data has continuous time intervals, facilitating subsequent data analysis, modeling, and visualization.
- Spatial interpolation is a method that infers the values at unknown locations through known discrete spatial data points. Its purpose is to estimate or predict the values at other locations based on the spatial relationships between the known data points (Wang et al. 2024b).

In air quality monitoring, both spatial interpolation and temporal interpolation have their unique characteristics and application scenarios.

- Temporal interpolation relies on the continuity and trend of time series data, utilizing known air quality data at specific time points to predict and estimate the air quality at a future time point. It can handle missing values in time series data, enhancing data completeness and availability. When monitoring data is insufficient or has low temporal resolution, temporal interpolation can be applied to improve the temporal resolution and prediction accuracy of the data.
- Spatial interpolation, based on the principle of geographical proximity and similarity, utilizes known air quality data from monitoring stations to predict and estimate air quality in unknown locations or unmonitored areas. It fills in data gaps and provides comprehensive air quality information. In cases where monitoring stations are insufficient or unevenly distributed, spatial interpolation can be used to predict the air quality in unmonitored areas.

7.2 Data Acquisition

Beijing is a heavily polluted city dominated by heavy industries and has seen significant improvement in recent years as a result of a series of vigorous measures taken by the government and concerted efforts made by the public.

The data comparisons in both Sects. 7.3 and 7.4 of this chapter are derived from monitoring data collected from 35 monitoring stations in Beijing shown in Fig. 7.1. The time interpolation comparison in Sect. 7.3 focuses on the average PM2.5 levels measured by these 35 stations over a consecutive 3-day period in Beijing. In

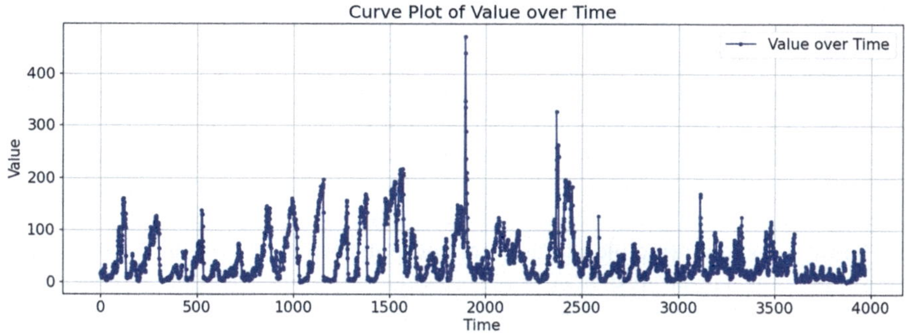

Fig. 7.1 The curve plots of the PM2.5 data in Beijing

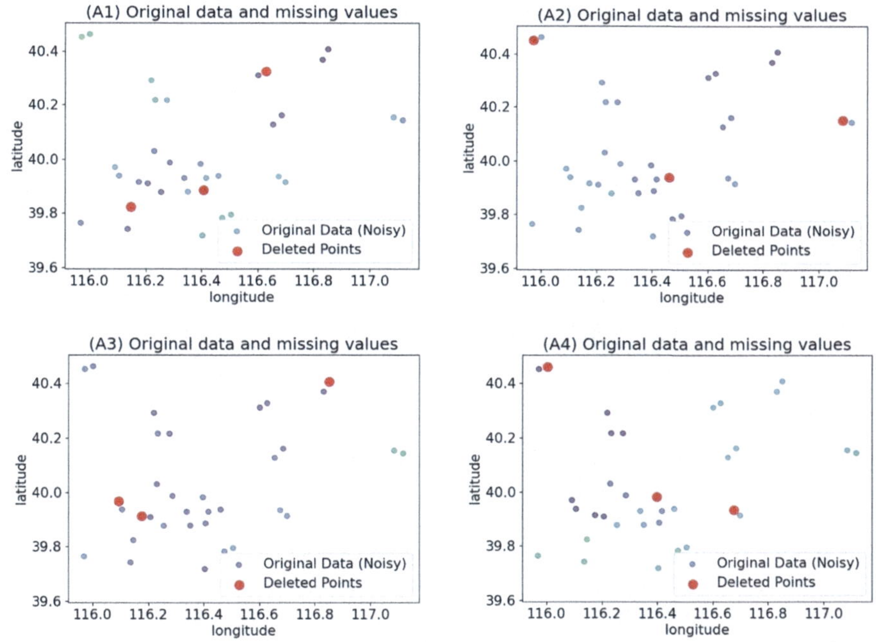

Fig. 7.2 The position diagram of the PM2.5 data of 35 stations in Beijing

contrast, Sect. 7.4 presents the PM2.5 levels at the same moment in time from the 35 monitoring stations in Beijing shown in Fig. 7.2, and spatial interpolation is performed based on the longitude and latitude coordinates of these 35 stations to estimate PM2.5 values at unmeasured locations. Four sets of data are the PM2.5 data monitored by 35 stations at different times in Beijing and randomly delete three sets of data for experimental comparison.

7.3 Temporal Interpolation of Air Quality Data

Temporal interpolation in the context of air quality data primarily involves the use of time-series data to estimate or predict air quality conditions at other time points based on observed values at known time points. This method is of significant importance in air quality monitoring and prediction, as it enables a more comprehensive understanding of the dynamic changes in air quality.

The fundamental principle of temporal interpolation primarily relies on the continuity and correlation present in time-series data. In time-series data, there often exists a certain degree of association between adjacent time points, meaning that the data value at one-time point may be influenced by the values at preceding or subsequent time points. Temporal interpolation leverages this association to estimate or predict data values at unknown time points based on known data. Specifically, the fundamental principles of temporal interpolation can be summarized as follows (Cai et al. 2021):

- Data Continuity Assumption: Temporal interpolation assumes that time-series data exhibits continuity over time, the data values change smoothly or follow a predictable pattern over time. This continuity allows us to infer unknown data points based on known data points.
- Correlation Analysis: Temporal interpolation also relies on the correlation within time-series data. Correlation analysis helps us understand the relationship between data values at different time points, thereby enabling the selection of appropriate interpolation methods to estimate unknown data. For instance, if the data exhibits a linear trend, linear interpolation can be used; for more complex data variations, more advanced interpolation methods such as polynomial interpolation, spline interpolation, or time-series prediction based on statistical models may be required.
- Interpolation Method Selection: Depending on the characteristics of the data and unique requirements, different temporal interpolation methods can be chosen. Commonly used temporal interpolation methods include linear interpolation, polynomial interpolation, spline interpolation, among others. Each method has its own merits and demerits, and the choice should be made based on the actual situation. For example, linear interpolation is easy and fast but may not accurately reflect nonlinear data changes; polynomial interpolation can fit more complex data variations but requires attention to overfitting issues; spline interpolation can better preserve local data features and is suitable for scenarios requiring high interpolation accuracy.
- Model Validation and Evaluation: After selecting and applying a temporal interpolation method, it is necessary to validate and evaluate the interpolation results. This typically involves comparing the interpolation results with known data to assess their accuracy and reliability. If significant deviations exist between the interpolation results and actual data, the interpolation method may need to be reselected or adjusted (Lepot et al. 2017).

In summary, the fundamental principle of temporal interpolation is to utilize the continuity and correlation of time-series data to estimate or predict data values at unknown time points based on known data. In practical applications, it is essential to select appropriate interpolation methods based on the characteristics and requirements of the data, and conduct model validation and evaluation to ensure the accuracy and reliability of the interpolation results. And four temporal interpolations were introduced in this section.

7.3.1 Linear Interpolation

7.3.1.1 Theoretical Basis

Linear Interpolation is a simple and widely used interpolation method that estimates the data value at an unknown time point between two known data points. Its fundamental principle assumes that the variation of data values between these two known points is linear over time, meaning the rate of change of data values remains constant. Specifically, linear interpolation involves connecting these two known data points to form a straight line (or segment), and then finding the corresponding data value on this line for the unknown time point. This process is analogous to measuring a position between two markings on a ruler and estimating the value at that position through proportionality (Dong et al. 2021).

The advantages of linear interpolation lie in its simplicity and speed of calculation, as well as its tendency to produce accurate estimates in scenarios where data changes relatively smoothly. Consequently, linear interpolation is widely employed in various fields such as meteorological observations, financial analysis, image processing, and many others. However, linear interpolation also has certain limitations. When data exhibits nonlinear changes, linear interpolation may fail to accurately reflect the true nature of the data, leading to deviations in the estimated results. In such cases, more complex interpolation methods, such as polynomial interpolation, spline interpolation, or time series forecasting based on statistical models/machine learning models, may need to be considered.

In summary, linear interpolation is a simple and effective interpolation method suitable for scenarios where data changes relatively smoothly. In practical applications, it is essential to select the appropriate interpolation method based on the characteristics of the data and the requirements of the scenario.

7.3.1.2 Modeling Step

To obtain the sum of averages of PM2.5 data from various monitoring stations in Beijing over three consecutive days and simulate it using linear interpolation, a comparison chart as shown in Fig. 7.3 is required. First, we need to identify two known data points, assuming that their relationship can be approximated by a

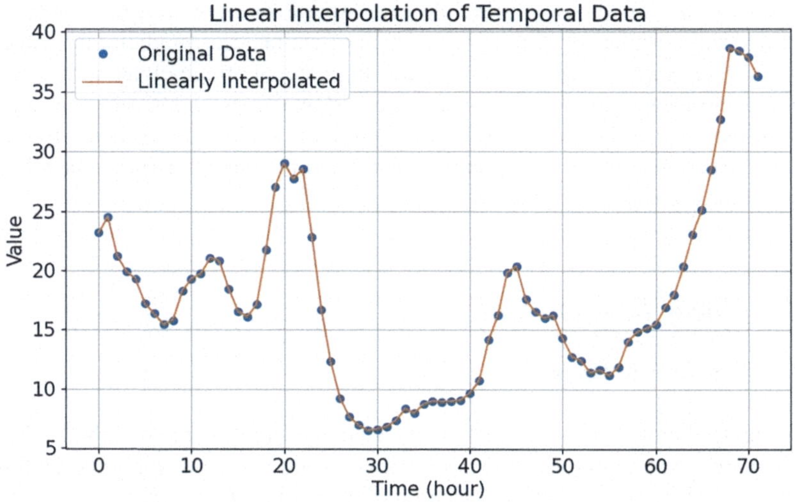

Fig. 7.3 The plots of Linear interpolation of the PM2.5 data in Beijing

straight line. Then, we calculate the slope K and determine the x-coordinate of the point that needs to be interpolated.

7.3.2 Polynomial Interpolation

7.3.2.1 Theoretical Basis

The principle of polynomial interpolation is to construct a polynomial function based on a set of discrete data points in time, and this function can accurately pass through these data points and interpolate between them to estimate or predict the values at other time points.

In time series data, we may encounter situations where data points are discontinuous or missing, and it is necessary to utilize the existing data points to estimate or fill in the missing time points. Polynomial interpolation is an effective tool to address this issue. It requires finding a polynomial function p(t), where t represents time, such that the function satisfies $p(t_i) = y_i$ at the given time points t_0, t_1, $t_2 \ldots t_n$, where y_i is the known data value corresponding to the time point t_i. The steps to construct a polynomial function are as follows (Ngoc et al. 2024):

Choosing the Degree of the Polynomial: Firstly, the degree of the polynomial needs to be determined. While a higher degree polynomial may more accurately approximate the original data, it can also lead to overfitting or the Runge's Phenomenon. Therefore, the appropriate degree of the polynomial should be selected based on the actual data situation.

Formulating the System of Equations: According to the definition of polynomial interpolation, a system of linear equations can be formulated to solve for the coefficients of the polynomial. Let the polynomial be $p(t) = a_0 + a_1t + a_2t^2 + \ldots a_nt^n$. For each known data point (t_i, y_i), we have $p(t_i) = y_i$. Combining these equations results in a system of linear equations involving the coefficients a_0, a_1, $a_2 \ldots a_n$.

Solving the System of Equations: Solving this system of linear equations yields the coefficients of the polynomial. Since it is a system of linear equations in terms of the coefficients, linear algebra methods such as Gaussian elimination or LU decomposition can be used to find the solutions.

After obtaining the coefficients of the polynomial, we can use the polynomial function $p(t)$ to perform interpolation calculations for other time points. For point t', simply substitute t' into the polynomial $p(t)$ to obtain the estimated value $p(t')$ at that time point.

7.3.2.2 Modeling Step

The degree of the polynomial interpolation (the highest power of the polynomial) is a crucial parameter that determines the complexity and fitting accuracy of the interpolating polynomial. Based on the given data points and the chosen degree of the polynomial, the Lagrange interpolation method can be employed to construct the interpolating polynomial. After constructing the polynomial, a graph depicting the polynomial curve alongside the data points can be plotted as shown in Fig. 7.4.

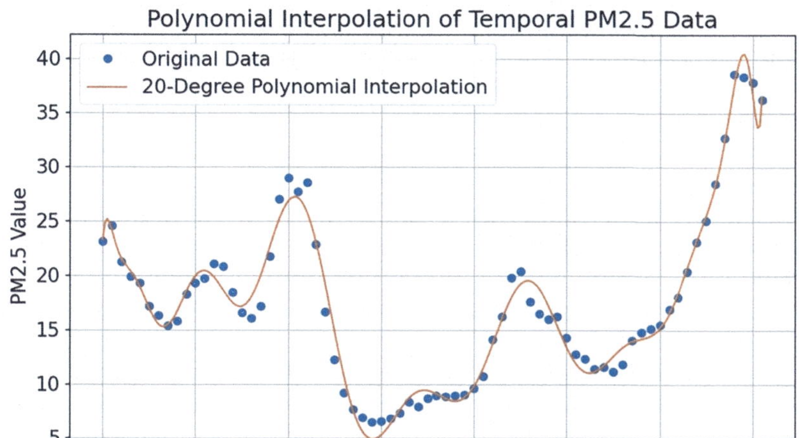

Fig. 7.4 The plots of Polynomial interpolation of the PM2.5 data in Beijing

7.3.3 Spline Interpolation

7.3.3.1 Theoretical Basis

Spline Interpolation is a mathematical method for constructing smooth curves through known data points. In temporal interpolation, Spline Interpolation is employed to estimate and interpolate unknown points in time-series data, resulting in a smooth and continuous time-series curve. The fundamental principles of Spline Interpolation lie in continuity and smoothness. Mathematically, this manifests as the equality of function values at junction points (continuity) and the equality of lower-order derivatives at those points (smoothness). Spline Interpolation typically utilizes piecewise polynomial functions, where each polynomial segment smoothly transitions between adjacent points.

Taking the most commonly used Cubic Spline Interpolation as an example, its formula is expressed as a series of cubic polynomial functions, each defined over an interval between adjacent data points. Assuming there are $n + 1$ data points (x_i, y_i) where $i = 0, 1, 2, \ldots, n$, the cubic polynomial on each interval $[x_i, x_{i+1}]$ can be represented as: $S_i(x) = a_i x^3 + b_i x^2 + c_i x + d_i$, where a_i, b_i, c_i, d_i are the coefficients that need to be solved. To determine these coefficients, the following conditions must be satisfied (Peng et al. 2024):

Interpolation Condition: Each polynomial equals the corresponding y value at the data points, i.e., $S_i(x_i) = y_i$ and $S_i(x_{i+1}) = y_{i+1}$

Continuity Condition: The function values of adjacent polynomials are equal at the joining points, i.e., $S_i(x_{i+1}) = S_{i+1}(x_i + 1)$

Smoothness Condition: The first derivatives (slopes) and second derivatives of adjacent polynomials are equal at the joining points, i.e., $S_i^{'}(x_{i+1}) = S_{i+1}^{'}(x_{i+1})$ and $S_i^{''}(x_{i+1}) = S_{i+1}^{''}(x_{i+1})$

These conditions collectively form a system of linear equations that need to be solved for the coefficients. Typically, in addition to the above conditions, boundary conditions are also introduced to fully determine the solution. The boundary conditions can be (Purwani et al. 2023):

Natural Boundary Conditions: where the second derivatives at the endpoints are zero (e.g., $S_0^{''}(x_0) = 0$ and $S_n^{''}(x_n) = 0$.

Clamped Boundary Conditions: where the first derivatives at the endpoints are known (e.g., $S_0^{'}(x_0) = m_0$ and $S_n^{'}(x_n) = m_0$ for some specified slopes m_0 and m_n

Non-knot Boundary Conditions: which impose additional constraints, such as specifying the third derivative of the spline curve to match certain conditions.

By solving this system of linear equations with the appropriate boundary conditions, the coefficients a_i, b_i, c_i, d_i can be determined, allowing for the construction of a smooth and continuous interpolating spline curve through the given data points.

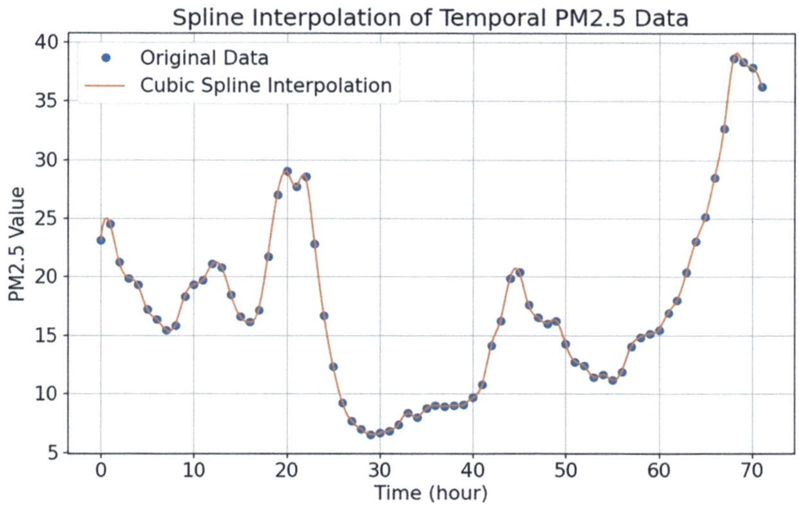

Fig. 7.5 The plots of Spline interpolation of the PM2.5 data in Beijing

7.3.3.2 Modeling Step

Cubic spline interpolation is widely used due to its smoothness and computational efficiency. This section also selects cubic spline interpolation as a representative method. Based on the selected spline type and interpolation conditions, the coefficients of the spline interpolation function are obtained by solving the corresponding mathematical equations. The original data points are plotted in the graphical interface or comparison with the interpolation results shown in Fig. 7.5. According to the constructed cubic spline interpolation function, a smooth spline curve is drawn between the data points.

7.3.4 Interpolation Based on Statistical Model

7.3.4.1 Theoretical Basis

The ARIMA model, short for Autoregressive Integrated Moving Average Model, is a widely used approach in time series analysis, especially effective for managing non-stationary data. The ARIMA model comprises three primary elements: Autoregressive, Integrated, and Moving Average, typically denoted as ARIMA.

Autoregressive: This represents a linear relationship between the current observation and a series of past observations. The autoregressive order defines the number of past observations considered in the model.

Integrated: This involves performing differencing operations on the time series, i.e., taking the difference between the current observation and the previous one. The differencing order indicates how many differencing operations are applied to make the time series stationary.

Moving Average: This signifies a linear relationship between the current observation and a series of white noise errors from past observations. The moving average order specifies the number of white noise errors considered in the model (Wu 2013).

The ARIMA model can indirectly achieve interpolation effects through the following steps (Kim and Kim 2021):

Data Preprocessing: First, examine the stationarity of the time series data. If the data is non-stationary, perform differencing operations until the data becomes stationary.

Model Selection and Parameter Estimation: Based on the stationary series obtained after differencing, select appropriate ARIMA model parameters. This is typically done by observing the autocorrelation and partial autocorrelation plots.

Model Fitting: Fit the ARIMA model using the selected parameters.

Model Diagnosis: Check if the residuals of the fitted model are white noise to ensure the model's validity.

Prediction and Interpolation: Utilize the fitted ARIMA model to forecast future time points. These forecasted values can be considered as "interpolated" estimates for unknown time points.

7.3.4.2 Modeling Step

Based on the characteristics of the data and the requirements of interpolation, an appropriate statistical model is selected for interpolation. Taking the ARIMA model as an example in this section, the preprocessed data is used to train the selected statistical model, and the model's parameters are solved through an optimization algorithm. Utilizing the trained statistical model, interpolation calculations are performed between the data points, and the interpolation results are obtained as shown in Fig. 7.6.

7.4 Spatial Interpolation of Air Quality Data

Spatial Interpolation plays a significant role in air quality data analysis, as it enables the transformation of discrete monitoring points' air quality data into a continuous data surface, providing a more comprehensive understanding of the distribution of air quality.

The core of spatial interpolation lies in utilizing the data from known points (air quality monitoring stations) to predict the data at unknown points. These known points, typically air quality monitoring stations positioned at specific locations, are capable of monitoring air quality parameters (such as PM2.5, PM10, sulfur dioxide, nitrogen oxides, etc.) in real-time or on a regular basis. By analyzing the spatial relationships and attribute value relationships between these known points,

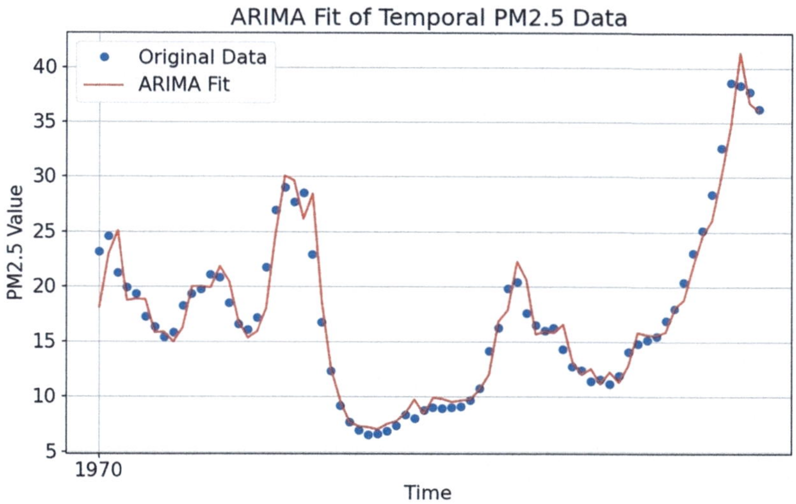

Fig. 7.6 The plots of interpolation based on the ARIMA Model of the PM2.5 data in Beijing

mathematical models can be established to estimate the air quality values at unknown points.

Spatial interpolation is extensively applied in air quality data analysis, as it facilitates a more comprehensive understanding of the distribution of air quality. Furthermore, spatial interpolation can be utilized for air quality data prediction and early warning. By integrating historical and real-time data, prediction models can be constructed using spatial interpolation methods to forecast air quality conditions over a future period. This enables timely warning information to be provided to the public and government, facilitating the adoption of corresponding response measures. In this section, four spatial interpolations were introduced.

7.4.1 Nearest Neighbor Interpolation

7.4.1.1 Theoretical Basis

In spatial interpolation, Nearest Neighbor Interpolation, also known as Nearest Neighbor Interpolation or Zero-order Interpolation, is a simple and intuitive interpolation method. The core idea of this method is that for an unknown interpolation point, the value of the known data point that is closest to it is directly adopted as the interpolation result for that point (Yi et al. 2022).

In spatial interpolation, the implementation steps of Nearest Neighbor Interpolation typically include the following aspects:

Determine the interpolation range: Clearly define the area within which interpolation is required.

Select known data points: Choose the known data points within the interpolation range that will be used for interpolation.

Calculate distances: For the unknown interpolation point, calculate the distances to all known data points.

Select the nearest point: From the known data points, select the one that is closest to the unknown interpolation point.

Assign the value: Assign the value of the selected nearest data point to the unknown interpolation point, using it as the interpolation result for that point.

7.4.1.2 Modeling Step

To monitor and interpolate the PM2.5 levels at 35 monitoring stations in Beijing during the same period, Nearest Neighbor Interpolation is used. For each interpolation point, find the point with the closest distance among the known data points. Assign the value of the found nearest neighbor point to the interpolation point. Further optimization or adjustment can be made to the interpolation results. A comparison between the original and the interpolated data is shown in Fig. 7.7.

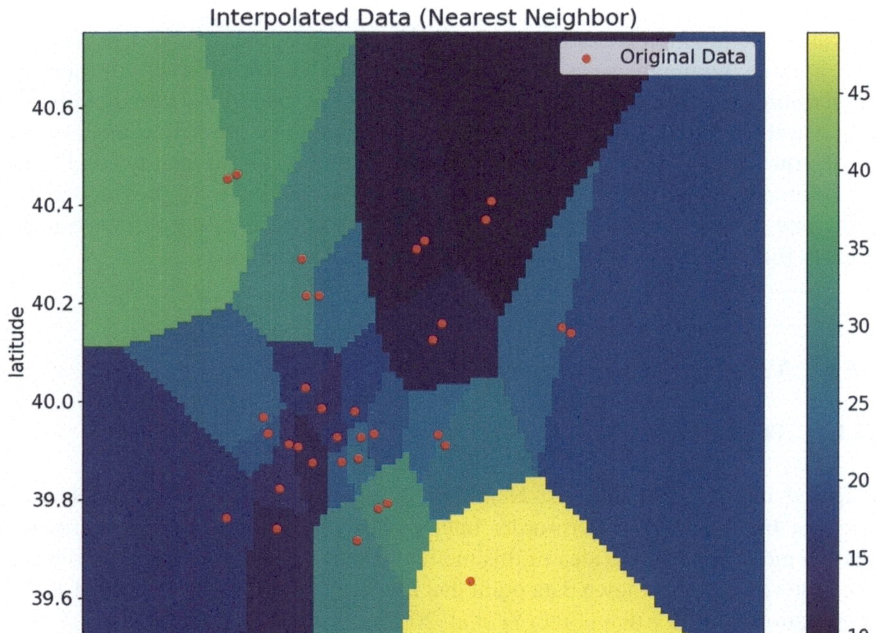

Fig. 7.7 The comparison diagram of Nearest Neighbor of the PM2.5 data in 35 stations in Beijing

7.4.2 *Inverse Distance Weighted Interpolation*

7.4.2.1 Theoretical Basis

Inverse Distance Weighted Interpolation (IDW) is a classical spatial interpolation method based on the fundamental assumption of Tobler's First Law of Geography, which states that "everything is related to everything else, but near things are more related than distant things." IDW interpolation estimates the value of an unknown point by considering the distances and values of surrounding known points. Its core principle lies in assigning larger weights to points closer in distance and smaller weights to points farther away. A detailed breakdown of the key components and the specific formula used in IDW are as follows:

The crux of DW (or IDW) interpolation is the calculation of weights, which are functions of distance, typically represented as the inverse power of distance. The specific formula is as follows:

Distance Calculation: For each known sample point and the interpolation point, the Euclidean distance is calculated. In two dimensions, this is given by: $d_i = \sqrt{(x_0 - x_i)^2 + (y_0 - y_i)^2}$ where $(x_0, y_0)$is the coordinate of the interpolation point and (x_i, y_i) is the coordinate of the i sample point (Zhang et al. 2024).

Weight Calculation: The weight $w_i$for each sample point is computed as the inverse power of the distance d_i: $w_i = \frac{1}{d_i^p}$, where p is the power parameter, a positive real number that controls the rate of weight decrease with distance. Higher values of $p$emphasize closer p points more strongly, while lower values distribute the influence more evenly among surrounding points.

Interpolated Value Calculation: The interpolated value v_0 at the interpolation point is computed as the weighted average of the values v_i of the surrounding sample points:

$$v_0 = \frac{\sum_{i=1}^{n} w_i \cdot v_i}{\sum_{i=1}^{n} w_i}$$

where n refers to the number of sample points used in the interpolation (Li et al. 2018).

7.4.2.2 Modeling Step

The IDW method assigns weights to observed points by computing the reciprocal of the power of the distance. When selecting a power value of 2 for the weights, a range is defined to specify the neighboring observed points used for interpolation. Based on the distance and the selected power value, the weight of each observed point is calculated. The weights are inversely proportional to the distance, and a

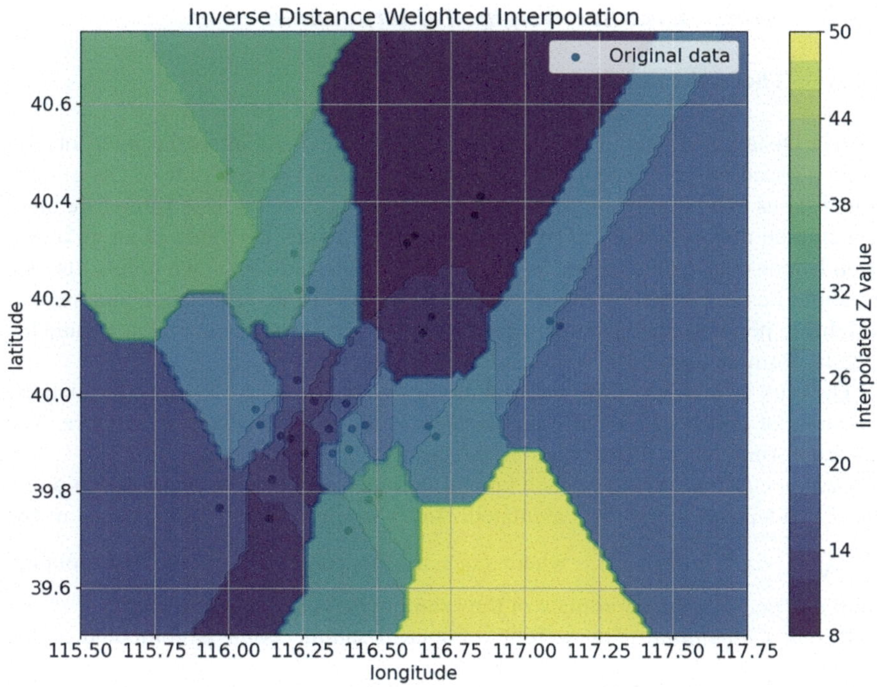

Fig. 7.8 The comparison diagram of IDW of the PM2.5 data in 35 stations in Beijing

weighted average approach is used to compute the attribute value at the interpolation point. Finally, a comparison between the original and the interpolated data is presented in Fig. 7.8.

7.4.3 Kriging Interpolation

7.4.3.1 Theoretical Basis

The principle of Kriging Interpolation, rooted primarily in statistics and geostatistics, is employed to estimate the unknown values of spatially continuous variables. It operates under the following key assumptions and concepts (Wang et al. 2024a):

Spatial Correlation: Kriging Interpolation assumes that the variable values in space possess a degree of continuity and correlation, implying that points closer in proximity tend to exhibit more similar attribute values. This correlation gradually diminishes as the distance between points increases.

Variogram: The quantification of the strength and extent of spatial correlation is achieved through the calculation of the variogram. The variogram serves as a core tool in Kriging Interpolation, describing the average degree of difference in attribute values across varying distances. It encapsulates the spatial variability and correlation patterns inherent in the data.

Optimal Unbiased Estimation: The objective of Kriging Interpolation is to provide an optimal unbiased estimate for the unknown points. This implies that the prospective value of the estimate equals the true value, and the variance of the estimation error is minimized. In other words, it aims to deliver the most accurate prediction possible, with the least amount of uncertainty.

7.4.3.2 Modeling Step

Kriging Interpolation takes into account the spatial relationships between sample points, utilizing variogram and structural analysis to provide an optimal unbiased estimate of the values at unknown points. It involves calculating the spatial distances between known points and the corresponding differences in attribute values. The variogram model is then fitted to the known data points, determining its parameters. Based on the attribute values, spatial coordinates of the known points, and the variogram model, spatial weights between unknown points and known points are computed. Finally, the interpolation results are visualized in the form of maps, charts, or other graphics. A comparison between the original and the interpolated data is presented in Fig. 7.9.

7.4.4 Radial Basis Function Interpolation

7.4.4.1 Theoretical Basis

Radial Basis Function (RBF) Interpolation is a widely employed mathematical method for spatial interpolation, particularly suitable for handling irregular data in high-dimensional spaces. Its fundamental concept lies in utilizing a set of kernel functions known as radial basis functions (RBFs) to construct a continuous spatial function from known discrete data points, enabling predictions and interpolations at unknown points. The principle overview is as follows (Perinajová et al. 2024):

Definition of the Interpolation Problem: Suppose we have a set of discrete data points $\{(x_i, f_i)\}_{i=1}^{N}$, where x_i represents a location in space and f_i is the known function value at that location. The objective is to find a function $s(x)$ that satisfies the following condition:$s(x_i) = f_i$, $\forall_i = 1, 2, \ldots, N$ and this function can be utilized to estimate the function values at any unknown point $s(x)$ in the space.

Radial Basis Functions: At the core of RBF interpolation lies the radial basis function $\phi(\|x - x_i\|)$, which depends solely on the Euclidean distance between the input point $\|x - x_i\|$ and the known data points. Commonly used RBFs include: the Gaussian Function: $\phi(r) = e^{-(er)^2}$ where r is the distance and ϵ is a scaling

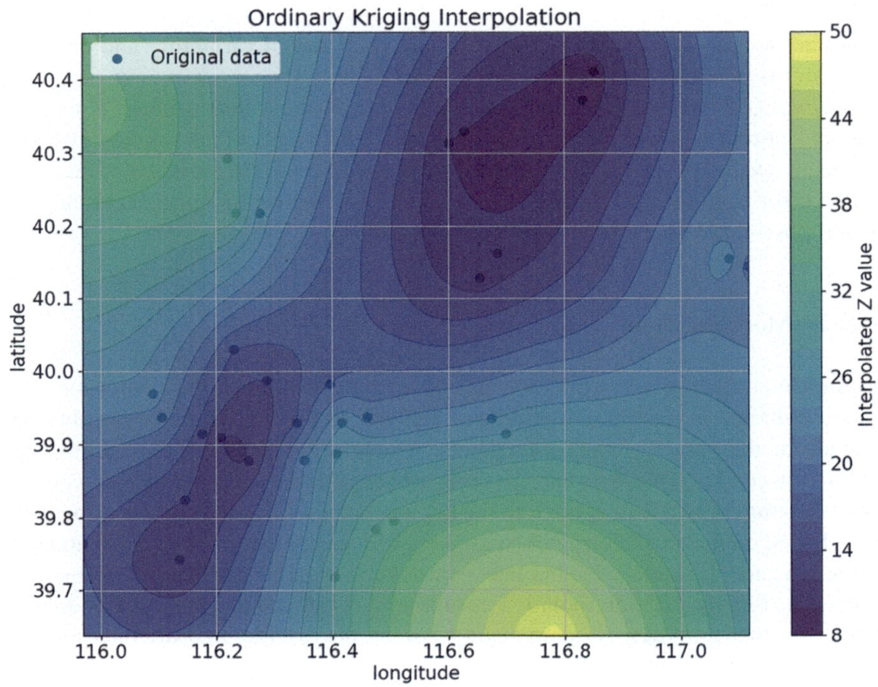

Fig. 7.9 The comparison diagram of Kriging interpolation of the PM2.5 data in 35 stations in Beijing

parameter. Multiquadric Function: $\phi(r) = \sqrt{r^2 + e^2}$. Thin Plate Spline:$\phi(r) = r^2 \log(r)$, These RBFs are crucial in defining the interpolation function as they spread from the discrete data points, naturally capturing local variations in the space (Amin and Voosoghi 2021).

Form of Interpolation Function: The RBF interpolation function $s(x)$ is typically constructed as a weighted sum of RBFs evaluated at all data points: $s(x) = \sum_{i=1}^{N} \lambda_i \phi(\| x - x_i \|)$ where λ_i are the weights to be determined, and $\phi(\|x - x_i\|)$ is the RBF based on the distance between the interpolation point and the data point.

Solving for Weights λ_i: To ensure interpolation accuracy, the interpolation function must match the known function values at each data point, f_i. This results in a linear system of equations: $s(x) = \sum_{i=1}^{N} \lambda_i \phi(\| x - x_i \|) = f_i, \forall i = 1, 2, \ldots, N$. In matrix form, this becomes: $\phi\lambda = f$ where the matrix ϕ has elements $\phi_{ij} = \phi(\|x - x_i\|)$, λ is the vector of unknown weights, and f is the vector of known function values. By solving this linear system, the weights λ_i are obtained, allowing the construction of the interpolation function $s(x)$.

Once the weights λ_i are determined, the interpolation function $s(x)$ can be used to estimate the value at any unknown point x. As the interpolation function is a weighted sum of radial basis functions, its properties ensure good smoothness and accuracy between the data points.

7.4.4.2 Modeling Step

RBF Interpolation is characterized by its ability to handle arbitrarily distributed scattered data. In RBF interpolation, it is necessary to select an appropriate radial basis function. For this section, the Gaussian function is chosen. Based on the specific application scenario and data characteristics, parameters for the RBF interpolation are determined. Using the selected RBF function and the known data points, the inverse matrix or pseudo-inverse matrix is solved. The interpolation results are then visualized as shown in Fig. 7.10, providing a more intuitive understanding of the data's spatial distribution patterns and variation characteristics.

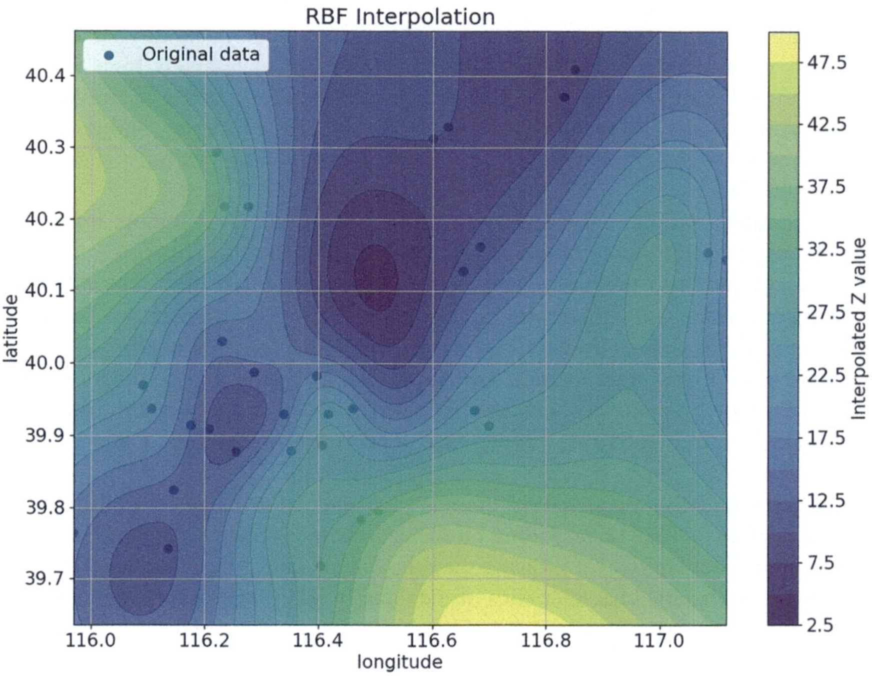

Fig. 7.10 The comparison diagram of RBF interpolation of the PM2.5 data in 35 stations in Beijing

7.5 Interpolation Performance Comparison

7.5.1 Comparison Between Temporal Interpolations

7.5.1.1 Characteristic of Four Temporal Interpolations

The advantages of Linear Interpolation are as follows, simple and easy to implement: The algorithm for linear interpolation is straightforward and does not require complex mathematical computations. Fast speed: Due to its simplicity, the computational speed is relatively fast. Insensitive to noise: Under low noise levels, linear interpolation can provide relatively stable interpolation results. However, Linear interpolation unable to capture nonlinear changes: For data with strong nonlinear trends, the accuracy of linear interpolation can be affected. Additionally, it is sensitive to outliers and the potential for overfitting: If there are outliers in the data, linear interpolation may produce significant errors. And when there are few data points and they are unevenly distributed, linear interpolation may generate interpolation results that do not conform to reality (Chen n.d.).

Polynomial Interpolation offers high flexibility by adjusting the degree of the polynomial to accommodate varying levels of data complexity, and it can provide high-precision interpolation results when sufficient and evenly distributed data points are available. However, it suffers from numerical instability, known as "Runge's phenomenon," when the polynomial degree is excessively high, and the computational complexity escalates significantly with the increase in polynomial degree. Additionally, polynomial interpolation is sensitive to noise, and data with significant noise levels may lead to distorted interpolation results (Ford and Quiring 2014).

Spline Interpolation boasts excellent smoothness, generating smooth interpolation curves, especially with numerous data points, and exhibits superior numerical stability compared to polynomial interpolation. It also offers high flexibility by allowing adjustments to spline order or boundary conditions to meet diverse interpolation requirements. However, spline interpolation is computationally more intensive than linear interpolation and low-order polynomial interpolation, and selecting optimal parameters such as spline order and boundary conditions can be challenging, often requiring experience and experimentation (Shatdal and Watzke n.d.).

Interpolation based on statistical models excels at capturing complex trends and periodic variations in data, providing more accurate interpolation results. It demonstrates strong adaptability by allowing the selection of appropriate statistical models tailored to the specific characteristics of the data. In addition to interpolation, these models possess robust predictive capabilities for future data values. However, the complexity of model selection necessitates expertise and experience, and the computational demands are higher compared to simpler interpolation methods. Furthermore, while statistical models generally exhibit some degree of noise resilience, excessively high noise levels can still adversely affect the interpolation outcomes.

7.5.1.2 Applications of Four Temporal Interpolations

When data changes are relatively smooth, with trends closely approximating linearity between data points, linear interpolation offers a simple yet effective means of interpolation. It is particularly favored in scenarios where computational resources are limited and a fast calculation is required, with precision not being the utmost priority, due to its straightforwardness. Furthermore, in cases where noise in the data is minimal or has a negligible impact on interpolation outcomes, linear interpolation can produce relatively stable interpolation results.

Polynomial Interpolation is applied when data changes are intricate, requiring high-precision interpolation. By fitting a polynomial curve, it effectively captures the complex trends between data points. This method excels when ample data points are available and evenly distributed, enabling it to deliver highly accurate interpolation results. However, it is crucial to be mindful of numerical stability, as excessively high polynomial degrees can lead to instability phenomena known as Runge's phenomenon, necessitating judicious selection of the polynomial degree in practical applications.

Spline Interpolation is employed when a smooth interpolation curve is desired, making it particularly suitable for scenarios where high smoothness of the interpolated data is crucial. With numerous data points, spline interpolation adeptly captures the trends while maintaining a smooth interpolation curve. Furthermore, its versatility allows for adjusting spline order or boundary conditions to cater to diverse interpolation requirements, demonstrating high flexibility in application.

Interpolation Based on Statistical Models is advantageous when dealing with complex data variations that necessitate both interpolation and prediction. It offers more precise interpolation results and the capability to forecast future data values. In situations involving vast amounts of data with diverse characteristics, selecting an appropriate statistical model tailored to the specific data properties enhances interpolation accuracy and adaptability. Additionally, while sensitive to noise, statistical models typically possess a degree of noise resilience, mitigating the impact of noise on interpolation outcomes to a certain extent.

7.5.1.3 Experiments Comparison of Temporal Interpolations

A comparative experiment was conducted on different interpolation methods. Firstly, the PM2.5 dataset from Beijing for 3 months was randomly selected and deleted, and then three interpolation methods, including linear interpolation, quadratic polynomial interpolation, and spline cubic interpolation, were applied. The diagram of time series with missing values after interpolation is shown in Fig. 7.11. By comparing the interpolated data with the original data and calculating Mean Squared Error (MSE), Root Mean Squared Error (RMSE), and Mean Absolute Error (MAE), the performance advantages of different models can be obtained given in Table 7.1.

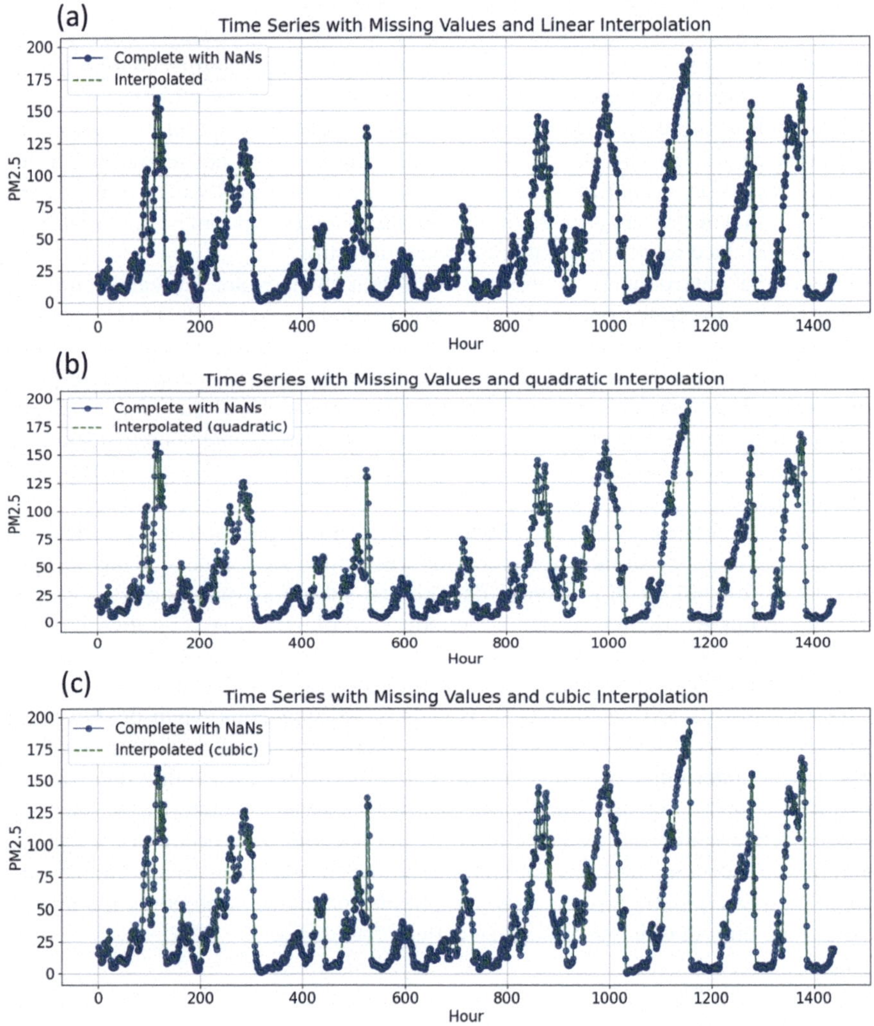

Fig. 7.11 Different interpolation diagram of time series with missing value. (**a**) Linear interpolation. (**b**) Quadratic interpolation. (**c**) Cubic interpolation

Table 7.1 The feature selection performance of different temporal interpolations

Types of temporal interpolation	RMSE (μg/m^3)	MAE (μg/m^3)
Linear interpolation	0.3546	0.0440
Quadratic interpolation	0.3226	0.0401
Spline cubic interpolation	0.3342	0.0428

As shown in Table 7.1, quadratic polynomial interpolation exhibits better performance due to the fact that when the data distribution is relatively uniform, without significant outliers or noise, especially when the data points are dense and the changes are relatively smooth, quadratic polynomial interpolation may be able to approximate the original data more accurately.

7.5.2 Comparison Between Spatial Interpolations

7.5.2.1 Characteristic of Four Spatial Interpolations

Nearest Neighbor Interpolation boasts the advantages of simplicity and speed, with a straightforward algorithm that enables rapid computation, making it suitable for real-time processing. Additionally, its ease of implementation, featuring straightforward and comprehensible code, renders it an ideal choice for beginners learning image processing or spatial interpolation. However, this method also suffers from drawbacks such as the noticeable jagged edges or "jaggedness" effect that can occur during image enlargement or spatial interpolation, compromising the smoothness of the interpolation results. Furthermore, it may lead to a loss of fine details, both in image reduction and spatial interpolation scenarios, potentially failing to accurately reflect subtle variations in the data (Yue and Shi 2025).

Inverse Distance Weighted Interpolation is a straightforward and efficient approach that adheres to the fundamental assumption of "Tobler's First Law of Geography," which means that everything is connected to everything else, but closer things are more closely connected than those farther apart. This method offers intuitive and effective interpolation, particularly when the known points are evenly distributed, resulting in interpolation outcomes that lie within the range of the data used for interpolation. However, it is susceptible to the influence of extreme values in the data, which can significantly affect the interpolation results. Furthermore, its high sensitivity to distances between sample points and interpolation points may lead to inaccurate interpolation outcomes in certain scenarios.

Kriging Interpolation stands out as a powerful tool that offers optimal unbiased estimation for unknown points, ensuring that the prospective value of the estimate equals the true value with minimal variance in the estimation error. It incorporates the spatial correlation between sample points, taking into account both their distances and relationships, thereby producing interpolation results that better align with actual spatial distribution patterns. Additionally, Kriging Interpolation boasts remarkable flexibility and adaptability, allowing for the selection of suitable interpolation methods and parameter settings based on the specific spatial distribution characteristics of the data and the form of the variogram. However, its computational complexity is higher than other interpolation methods, requiring more computational resources. Moreover, the choice of variogram model and parameter settings significantly impact the interpolation results, necessitating a degree of professional knowledge and experience (Guidoum 2025).

Radial Basis Function Interpolation excels in its local properties, delivering high fitting accuracy near data points, making it well-suited for scenarios requiring precise local interpolation. Its adaptability is remarkable, and it is capable of handling various data distributions with ease. Furthermore, by selecting appropriate radial basis functions and weight coefficients, a relatively simple interpolation model can be constructed, facilitating straightforward calculations. However, the interpolation effect is highly sensitive to the choice of radial basis functions, necessitating a problem-specific selection. Additionally, its robustness is somewhat limited, as it can be susceptible to noise, potentially leading to unstable interpolation results.

7.5.2.2 Applications of Four Spatial Interpolations

Nearest Neighbor Interpolation (NNI): Rapid Monitoring: When there is a need to quickly obtain interpolated results of monitoring data, NNI is suitable due to its simplicity and fast computational speed. However, this method may result in noticeable jagged edges, known as the "jaggedness" effect, in the interpolation results, affecting the smoothness of the data. Sparse Data Point Distribution: In situations where monitoring points are sparsely distributed, NNI can serve as a temporary or quick interpolation method, but caution should be taken regarding its potential impact on data smoothness (Antal et al. 2021).

When monitoring points are relatively evenly distributed within the monitored area, Inverse Distance Weighted Interpolation (IDW) effectively captures the spatial trends in monitoring data, providing accurate interpolation results. In environmental monitoring applications, such as air quality and water quality assessments, where estimations of environmental quality at unmonitored locations are required, IDW is frequently utilized due to its simplicity, intuitive nature, and high efficiency.

When high precision is required in air quality monitoring, Kriging interpolation is widely employed due to its ability to provide optimal unbiased estimates. It thoroughly considers the spatial relationships between monitoring points, resulting in more accurate interpolation outcomes. In cases where monitoring data exhibits complex spatial distributions, such as the presence of multiple pollution sources or significant terrain variations, Kriging interpolation effectively captures the spatial variability and correlation within the data, offering more reliable interpolation results.

Radial Basis Function Interpolation is applicable in scenarios requiring high-precision local interpolation, such as estimating monitoring data in specific areas near pollution sources, where it offers superior fitting accuracy to meet the monitoring needs of localized regions. Additionally, when monitoring data exhibits nonlinear spatial distribution, RBFI effectively approximates the data trends, resulting in more accurate interpolation outcomes. However, it is crucial to note that RBFI's sensitivity to the choice of radial basis functions necessitates careful selection based on the actual data (Xiao et al. 2025).

7.5.2.3 Experiments Comparison of Spatial Interpolations

This section discusses the comparison of experimental results for different spatial interpolation methods. The data consists of PM2.5 values at noon (12:00) on various dates from 35 stations in Beijing, labeled as A1, A2, A3, and A4. Data from relevant coordinate points are randomly selected and deleted, and the performance of different interpolation methods is compared.

The performance results of linear interpolation, IDW interpolation, Kriging interpolation, RBF interpolation on different datasets are shown in Figs. 7.12, 7.13, 7.14, and 7.15, respectively. In these figures, the horizontal axis represents longitude, the vertical axis represents latitude, and the data values at different coordinate points are marked accordingly. The data that has been selected and deleted is marked in red. By comparing the interpolated data calculated using these methods with the original accurate data, the performance of different interpolation methods can be demonstrated. And the performance of different spatial interpolation methods on datasets A1, A2, A3, and A4 is presented in Table 7.2.

From Table 7.2, among the four sets of data A1, A2, A3, and A4, the performance of the four interpolation methods varies, depending on the correlation between the

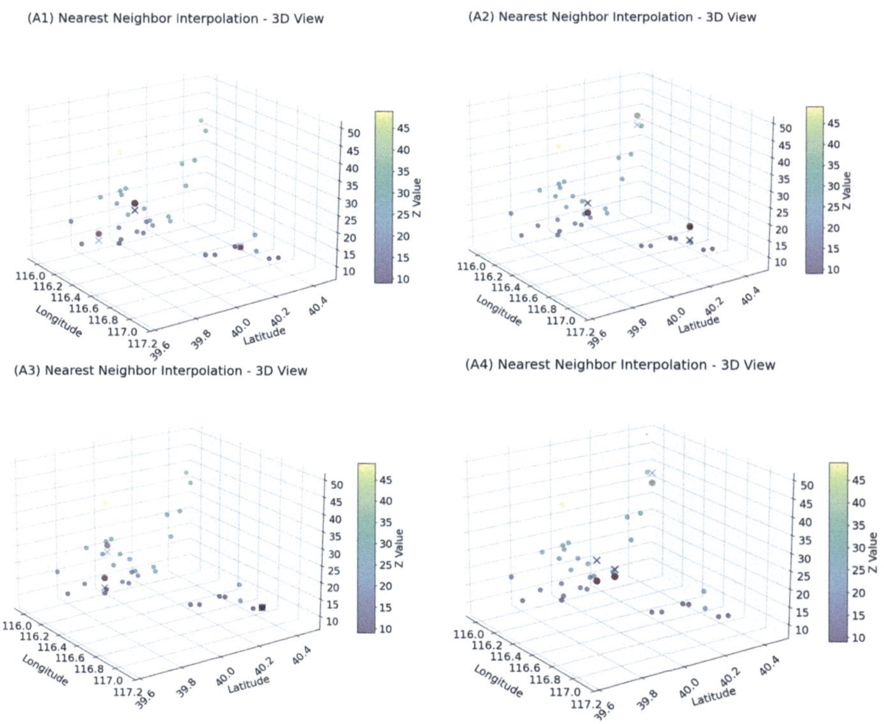

Fig. 7.12 Nearest interpolation diagram of different time series with missing value

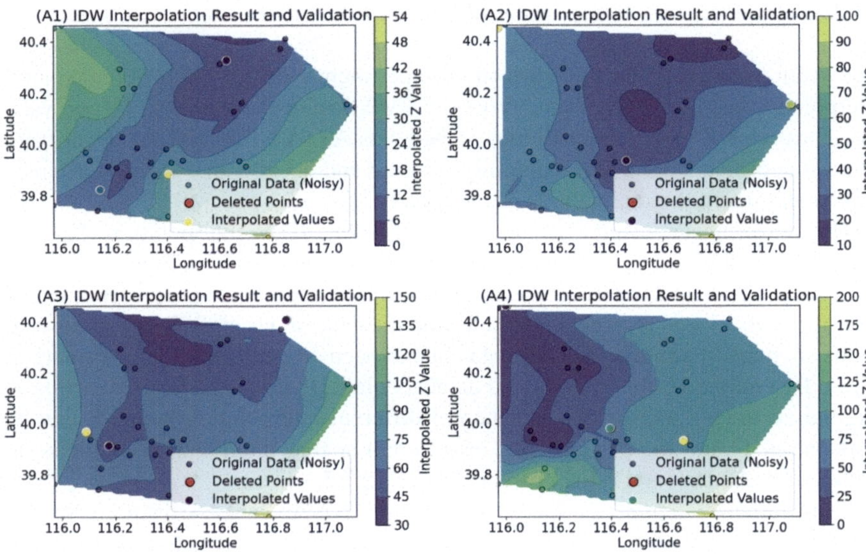

Fig. 7.13 IDW interpolation diagram of different time series with missing value

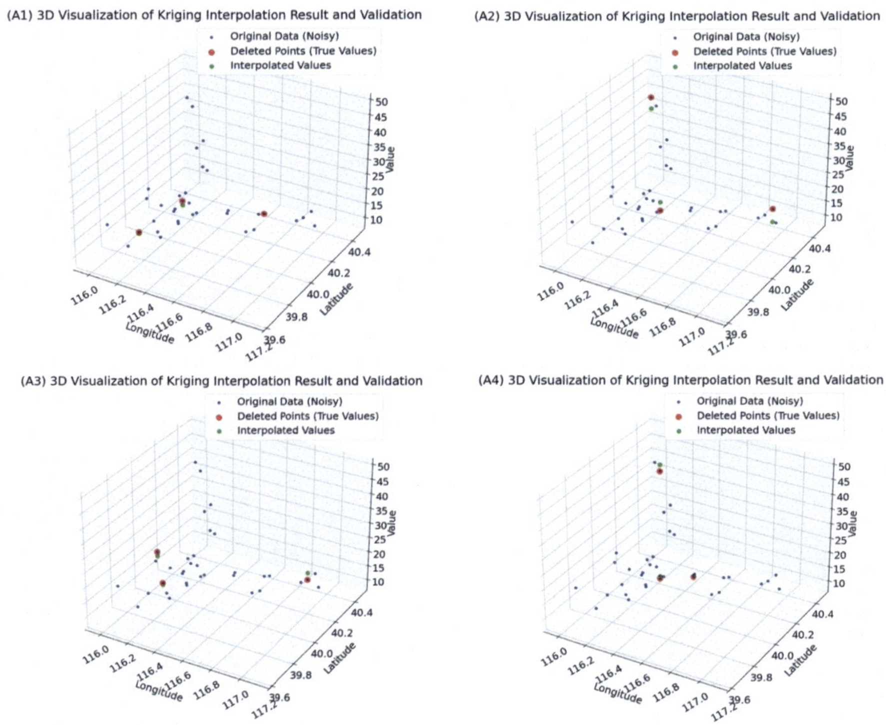

Fig. 7.14 Kriging interpolation diagram of different time series with missing value

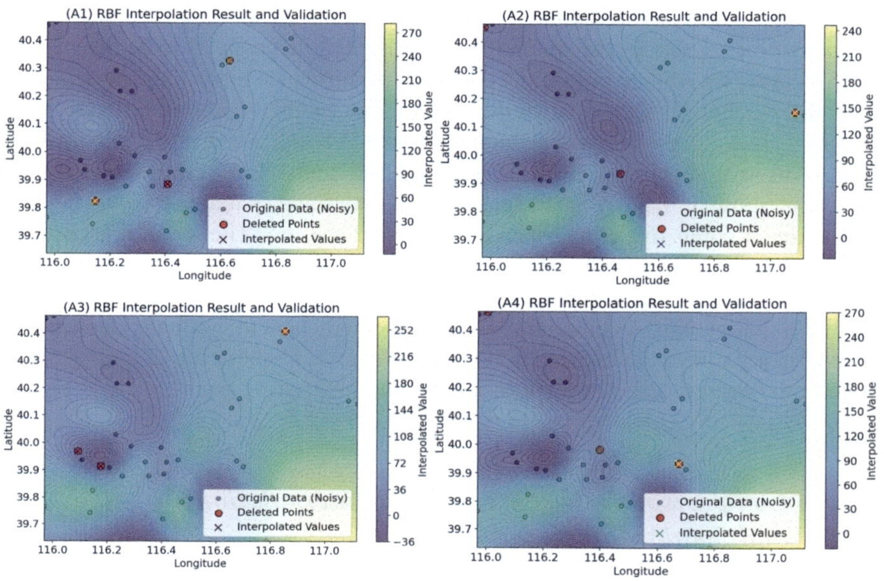

Fig. 7.15 RBF interpolation diagram of different time series with missing value

Table 7.2 The feature selection performance of different spatial interpolations

Types	Series	RMSE (μg/m³)	MAE (μg/m³)	MAPE (%)
Nearest neighbor	A1	1.63	1.33	7.33
	A2	2.71	2.67	7.87
	A3	13.11	11.33	22.17
	A4	5.23	3.33	4.24
IDW	A1	2.18	1.58	7.36
	A2	2.79	2.73	8.12
	A3	12.30	10.32	20.43
	A4	7.89	6.46	10.97
Kriging	A1	0.89	0.66	3.09
	A2	1.71	1.45	3.88
	A3	11.89	8.83	17.98
	A4	11.29	10.77	33.50
RBF	A1	1.60	1.35	7.33
	A2	3.75	2.59	9.32
	A3	18.15	14.59	29.01
	A4	10.54	8.79	15.20

deleted data points and the original data. The locations of the 35 coordinate points do not exhibit regularity and are unevenly distributed. If the point to be detected is far from other coordinate points, the detection performance tends to be poor.

In the A1 and A2 arrays, all four interpolation methods performed well, with Kriging achieving the highest accuracy in both A1 and A2 arrays, with MAPE (Mean Absolute Percentage Error) values of 3.09% and 3.88% respectively. This is because Kriging utilizes the statistical characteristics and spatial autocorrelation of measurement points to estimate data at unknown locations. Therefore, it is particularly suitable for data with significant spatial autocorrelation.

In the A4 array data, Nearest Neighbor exhibited a higher accuracy rate due to the dispersion of data. As other interpolation methods tend to gradually decrease in prediction accuracy with increasing distance, Nearest Neighbor directly assigns the value of the nearest detected point to the missing monitoring point.

7.5.3 Comparison Between Temporal Interpolation and Spatial Interpolation

In air quality monitoring, Temporal Interpolation and Spatial Interpolation are two commonly used data processing methods, each with distinct characteristics and applicable scenarios.

Temporal Interpolation primarily deals with time-series data, filling in unknown data points between known points to generate a smooth time-series curve. In air quality monitoring, it is often utilized to estimate missing hourly or daily average air quality data. The characteristics and advantages are as follows (Climate Research 2018):

Continuity: It enables the generation of continuous time-series data, facilitating the analysis of air quality trends over time.

Smoothness: By employing suitable interpolation methods, it can reduce noise and outliers in the time-series data.

Predictive Capability: In some cases, Temporal Interpolation can also be leveraged for short-term air quality forecasting.

Applicable Scenarios: When evaluating air quality in regions without monitoring stations. During regional air quality planning and management, when understanding the overall air quality distribution across the region is necessary.

Spatial Interpolation, on the other hand, estimates data for unknown spatial locations based on known spatial data points. In air quality monitoring, it is commonly used to assess air quality in areas without monitoring stations. The characteristics and advantages are shown in Table 7.3:

Regionality: It effectively captures the spatial distribution characteristics of air quality.

Diversity: A variety of interpolation methods are available, such as Inverse Distance Weighted and Kriging, each suitable for different data distribution patterns.

Table 7.3 The comparison between temporal interpolation and spatial interpolation

Characteristic	Temporal interpolation	Spatial interpolation
Objective	Filling in missing data within time series	Estimating data at unknown spatial locations
Data type	Time-series data	Spatial distribution data
Application scenarios	Analysis of air quality time series, short-term forecasting	Assessment of regional air quality, planning and management
Advantages	Continuity, smoothness, predictive capability	Regionality, diversity, high accuracy

High Accuracy: When data points are densely and evenly distributed, Spatial Interpolation can provide highly accurate interpolation results.

Applicable Scenarios: When there are temporal gaps in the data from air quality monitoring stations. When analyzing the long-term trends of air quality time-series data.

7.6 Conclusions

This chapter mainly discusses the application of data interpolation in air quality monitoring. Air quality monitoring is limited by the deployment density and measurement frequency of monitoring equipment, resulting in certain interruptions and gaps in the monitoring data both temporally and spatially. Data interpolation techniques are widely used to fill these temporal and spatial data gaps to assess air quality more accurately. This chapter mainly explores interpolation methods from two categories: temporal interpolation and spatial interpolation.

In the third and fourth parts, the main principles of temporal and spatial interpolation are illustrated and several typical temporal and spatial interpolation methods are introduced. Temporal interpolation estimates the missing data at specific time points for monitoring stations, allowing for a more continuous analysis of air pollution trends. Common temporal interpolation methods include linear, polynomial, spline, and interpolation based on statistical models. Spatial interpolation estimates spatial data between different monitoring stations to provide a more continuous assessment of regional air quality. Common spatial interpolation methods include nearest neighbor interpolation, inverse distance weighting interpolation, Kriging interpolation, and radial basis function interpolation.

Then, the fifth part of this chapter presents a comparison of the two major categories of interpolation methods. Firstly, performance experiments were conducted using 72-h PM2.5 data from Beijing on four temporal interpolation methods. Among the temporal interpolation methods, linear interpolation is simple but performs inadequately for larger data gaps; polynomial interpolation can better fit data trends but fluctuates significantly at the boundaries; spline interpolation exhibits remarkable smoothing effects and is suitable for more stable data sequences; while interpolation methods based on statistical models perform superiorly when data features are prominent. Subsequently, comparative experiments were conducted on four spatial interpolation methods using PM2.5 data from 35 monitoring stations in Beijing. In terms of spatial interpolation, nearest neighbor interpolation is simple but has limited effectiveness; inverse distance weighting interpolation can reflect spatial gradients but is susceptible to noise; Kriging interpolation has high accuracy and is suitable for air quality monitoring in complex terrains; radial basis function interpolation performs well when data distribution is sparse, but the computational load is relatively high.

References

Amin MK, Voosoghi B (2021) Gaussian radial basis function interpolation in vertical deformation analysis. Geod Geodyn 12(3):218–228

Antal A, Guerreiro PMP, Cheval S (2021) Comparison of spatial interpolation methods for estimating the precipitation distribution in Portugal. Theor Appl Climatol 145(3–4):1–14

Cai B, Shi Z, Zhao J (2021) Novel spatial and temporal interpolation algorithms based on extended field intensity model with applications for sparse AQI. Multimed Tools Appl 81(14):1–22

Chen W (n.d.) Method for constructing interpolation image frames in temporal interval, involves computing interpolated motion field at temporal interpolation time using conservative motion equation system. US2012147263-A1; US8559763-B2

Climate Research (2018) Reports on climate research findings from University of Texas provide new insights (A test of emergent constraints on cloud feedback and climate sensitivity using a calibrated single-model ensemble). Glob Warm. Focus 116

Dong Z, Wu C, Fu X, Wang F (2021) Research and application of back propagation neural network-based linear constrained optimization method. IEEE Access 9:126579–126594

Ford TW, Quiring SM (2014) Comparison and application of multiple methods for temporal interpolation of daily soil moisture. Int J Climatol 34(8):2604–2621

Guidoum A (2025) Statistical modeling and mapping of rainfall in the endorheic basins of Northern Algeria: a comparison of spatial interpolation methods. Acta Geophys 73:1679–1699

Kim J, Kim Y (2021) Statistical interpolation method for water quality data to improve water quality calibration and validation in watershed models. Ecohydrol Hydrobiol 21:67–78

Lepot M, Aubin J-B, Clemens FHLR (2017) Interpolation in time series: an introductive overview of existing methods, their performance criteria and uncertainty assessment. Water 9(10):796–796

Li Z, Wang K, Ma H, Wu Y (2018) An adjusted inverse distance weighted spatial interpolation method. In: 2018 3rd international conference on communications, information management and network security (CIMNS2018), p 5

Liu X, Zhang M, Zhang J, Yang Z, Gou Q, Deng S (2018) Application of an interpolation method in pollution survey by Matlab. IOP Conf Ser Earth Environ Sci 170(3):032051

Ngoc NA, Van Khiem N, Van Long T, Van Manh P (2024) Multivariate polynomial interpolation based on Radon projections. Numer Algor, 1–23

Peng W, Tao D, Ma Q, Xie Q, Wang J (2024) Clipped seismic record recovery analysis based on the cubic spline interpolation algorithm. J Seismol 28(3):843–857

Perinajová R, van de Ven T, Roelse E, Xu F, Juffermans J, Westenberg J, Lamb H, Kenjereš S (2024) A comprehensive MRI-based computational model of blood flow in compliant aorta using radial basis function interpolation. Biomed Eng Online 23(1):69

Purwani S, Hidayana RA, Balqis VP, Sukono (2023) A comparison of newton's divided differences interpolation and a cubic spline in predicting the poverty rate of West Java. Eng Lett 31(4):1786–1791

Roszkowiak L, Korzynska A, Zak J, Pijanowska D, Swiderska-Chadaj Z, Markiewicz T (2017) Survey: interpolation methods for whole slide image processing. J Microsc 265(2):148–158

Shatdal A, Watzke MW (n.d.) Data interpolation e.g. temporal interpolation, performing method for parallel database system, involves receiving database query with clause specifying time duration over which temporal interpolation is to be performed. US7236971-B1

Wang X, Ye P, Deng Y, Yuan Y, Zhu Y, Ni H (2023) Influence of different data interpolation methods for sparse data on the construction accuracy of electric bus driving cycle. Electronics 12(6):1377–1377

Wang J, Shi T, Wang H, Li M, Zhang X, Huang L (2024a) Estimating the amount of the WildArtemisia annuain China based on the MaxEnt model and spatio-temporal Kriging interpolation. Plants 13(7):1050

Wang Y, Liu X, Liu R, Zhang Z (2024b) Research progress on spatiotemporal interpolation methods for meteorological elements. Water 16(6):818

Wu J (2013) Improved application of K-N smooth linear interpolation method in measurement of English readability based on statistical language model. Int J Appl Math Stat 44(14):54–63

Xiao NB, Guo JS, Kuang ZJ, Wang W (2025) Application of data partitioned Kriging algorithm with GPU acceleration in real-time and refined reconstruction of three-dimensional radiation fields. Ann Nucl Energ 212:111047. https://doi.org/10.1016/j.anucene.2024.111047

Yi J, Tan H, Zhang J, Li Z (2022) Method of voltage setting for power battery simulator using successive nearest-neighbor interpolation. Meas Control 55(5–6):288–295

Yue ZJ, Shi MJ (2025) Enhancing space-time video super-resolution via spatial-temporal feature interaction. Neural Netw 184:107033. https://doi.org/10.1016/j.neunet.2024.107033

Zhang B, Cao J, Lin S, Li X, Zhang Y, Zheng X, Chen W, Song Y (2024) Optimized inverse distance weighted interpolation algorithm for γ radiation field reconstruction. Nucl Eng Technol 56(1):160–166

The manufacturer's authorised representative in the EU is Springer Nature Customer Service Centre GmbH, Europaplatz 3, 69115 Heidelberg, Germany. If you have any concerns regarding our products, please contact ProductSafety@springernature.com

Printed and bound by CPI Group (UK) Ltd, Croydon, CR0 4YY
01/10/2026
02225850-0008